THE UNINVITED II: THE VISITATION

David was still scanning the sky for the glowing object, thinking it would appear again at any second.

It didn't.

He exhaled, relaxing his grip on the steering wheel. A smile even touched his lips. 'Well, everybody, it looks like our excitement is over for the night,' he said. 'We should be home soon.'

He smiled to himself, trying to dismiss what had just happened as nothing more than 'a little adventure'. He shook his head. How empty and unconvincing those words sounded. Who was he trying to fool? He had been frightened as, he imagined, anyone in a similar position would have been. The sight of the unidentified . . .

He allowed the thought to trail off.

'Dad.'

'Yes, Dean,' he said.

'That light we saw.'

'What about the light?' he said.

'It's behind us.'

THE UNINVITED II: THE VISITATION

Frank Taylor

A STAR BOOK

published by
the Paperback Division of
W. H. ALLEN & Co. PLC

A Star Book
Published in 1984
by the Paperback Division of
W. H. Allen & Co. PLC
44 Hill Street, London W1X 8LB
Reprinted 1984

Copyright © 1984 Frank Taylor

Typeset by Phoenix Photosetting, Chatham
Printed in Great Britain by
Hunt Barnard Printing Ltd., Aylesbury, Bucks.

ISBN 0 352 31539 3

This book is sold subject to the condition that it shall not, by way of trade or otherwise, be lent, re-sold, hired out or otherwise circulated without the publisher's prior consent in any form of binding or cover other than that in which it is published and without a similar condition including this condition being imposed on the subsequent purchaser.

To Ray Mudie
When they ask you to blow in their bag,
it's not because their chips are hot.
Cheers.

'Fifty billion planets,
there's got to be some life.
Don't tell me I'm the only one
standing in the night . . .'

Saxon

Prologue

Every single incident described in this book is true. It actually happened. It happened to ordinary people who found themselves involved with something terrifying. Something unwanted, unexplained. Uninvited.

This book is not a work of fiction, although there are those who will try to dismiss it as such. But for every sceptic there will be a believer and it does not, in the final analysis, matter if this story is believed or not. For what you will read *did* happen. That is fact and nothing can be done to change it. Set down on the following pages is a series of occurrences for which there is no easy scientific explanation. Science, for once, is inadequate for finding answers to questions we scarcely dare ask. Nevertheless, the proof is here to be read.

The story which follows is not an isolated one. In 1977, the Coombs family in Wales were the subject of a similar, though less malevolent, visitation. Their story was told in *The Uninvited*, a book which was labelled by some as the product of an over-active imagination. It may well be that this book will be similarly criticised. That is unimportant. The facts speak for themselves and it is for you, as readers, to decide whether to believe or not. Whichever you decide, just be thankful that what you are about to read did not happen to you. Although it would be naïve to suggest that the events set down in this book will be the last of their kind.

There is too much in life which we don't understand. Too many questions to which we will never know the answers.

Perhaps, in cases such as that which follows, it is best not to know.

One

July 22nd

It was too bright to be a star.

David Ellis glanced out of his side window and studied the distant object for a second longer.

It was too big to be a star.

He focused his attention back on the road ahead, guiding the Volvo along at a steady forty miles an hour.

David had first noticed the object about ten minutes earlier, shortly after they had left the outskirts of Chesterfield behind. As he drove, surprised at how little traffic there was on the A632, he had spotted the glowing yellow light quite by chance. An aeroplane, he'd thought at first, but then he had decided that the object could well be a meteor or comet.

But a star?

He glanced at it again.

The bloody thing was getting bigger, he was certain of it.

David squeezed his eyes shut tightly for a second. He was tired, that was the answer, his eyes were playing tricks on him. The object wasn't getting bigger and, when he looked back again, if he bothered to, it would probably be gone. He decided to concentrate on his driving.

Beside him, his wife Caroline dozed. Her head lolled back against the head-rest. Her fingers were moving in an erratic rhythm, attempting to keep time with the tune which floated from the car radio. In the back seat their children, Michelle, who was eleven, and Dean, yet to reach his tenth birthday, slept. Caroline's sister, Julie, yawned and rubbed her eyes. She had lived with the family for the past three months, since the break-up of her marriage. At thirty-one, she was three years younger than Caroline. Five years David's junior. The two women were very much alike both physically and emotionally.

'David, what *is* that?'

The voice came from behind him and he turned slightly as Julie leant forward, pointing out of the side window.

She was indicating the yellow object.

'It's probably a star,' he said, trying to justify his own theory.

'I noticed it about five minutes ago,' Julie said.

David didn't speak but, instead, looked out of his side window again, squinting at the object.

It was easy to see, not only because of its glowing phosphorescence but because the night sky was so clear. The clock on the dashboard showed 10.36 p.m. and there wasn't a cloud to be seen in the dark blue canopy above them. Silvery pinpricks of stars hung in the heavens like sequins on velvet and, amongst them all, the yellow object shone out clearly. David frowned, realizing that its size had something to do with how low it was in the sky. He could have sworn that it had not been as low as that when he'd first seen it.

He sighed. It *must* be an aeroplane. And yet the lights which he saw had no break in them, the object seemed to be composed entirely of one single luminous entity.

The object, which he could see was now beginning to take on some discernible shape, passed over Walton Wood on their left, now, apparently, keeping pace with the car.

He smiled to himself. He felt so stupid. At first he had thought it was his eyes that were playing tricks on him, now he thought it was his mind. Perhaps the fact that he had seen that blasted film *Close Encounters of the Third Kind* so recently was helping to fuel his interest in this mysterious object. A spacecraft. He almost laughed aloud.

From her vantage point in the rear, Julie was watching even more intently. The object, from being a blur of yellow light, had now taken on a recognisable shape. It reminded her of a large egg.

David looked up again.

He swallowed hard. The object *was* getting lower in the sky, it was drawing closer.

They drove past Harper Hill, about six miles from the point where the main road was joined by the B5057, but David knew that he must keep to the main road and, for some unaccountable reason, that knowledge made him feel more secure. He glanced at the object once more. It was glowing much more brightly now, so much so that he had to narrow his eyes to look at it. He gripped the steering wheel tighter and drove on, guessing that it would take them perhaps another forty or fifty minutes to reach Matlock and home. As he glanced at the yellow shape again he blinked hard, his eyes feeling as if they were full of grit. He cursed under his breath and pushed his foot down slightly on the accelerator.

The light, burning ever brighter, dropped lower, picking up speed. It was directly in line with the car now.

Was this, David wondered, some kind of optical illusion? It was easy to imagine that something was following you. That the stars were moving. Perhaps this was the same sort of thing. He couldn't be sure how far away the glowing object was, how high in the sky. All he knew was that it seemed too close for comfort.

There was a deafening blast of static from the radio and Caroline sat bolt upright in her seat.

'What was that?' she said, touching her temple with one hand.

No one spoke. Then she, too, saw the object.

'David . . .' She was pointing at it.

He cut her short.

'I know, I've seen it,' he said, tersely.

She turned in her seat to see that Julie was gazing at the yellow egg and also that the children had been awoken by the deafening burst from the radio. There was more static and then silence. David twisted the frequency knob but there was no sound to be heard.

He pulled his hand away sharply.

The radio felt hot. Indeed, the entire dashboard was

beginning to give off a faint blanket of heat. The petrol gauge needle began to rise and fall slowly.

Outside, the glowing object was still tracking the car, now drawing closer until David guessed it was little more than a mile away from them, cutting through the sky like a missile. It left no vapour trail as an aircraft would but then, he reasoned, no aircraft would be flying so low. The Derbyshire countryside with its countless hills and valleys was no place for a pilot to practise low altitude flying. And this craft, whatever the hell it was, seemed to be less than seven hundred feet up by now.

'What is it, David?' asked Caroline.

'I don't know,' he said, quietly, his voice taut.

He tried to look away from the yellow thing, to concentrate his mind on the road and countryside ahead, perhaps in the hope that if he ignored it, the yellow beacon would disappear. He glanced at his petrol gauge again. The needle was still moving up and down, though more rapidly now.

He looked at the clock on the dashboard, the hands frozen at 10.49. He tapped it with his index finger then glanced at his own watch.

David felt a shiver run down his spine.

He held the watch to his ear but could hear no ticking. It had stopped at 10.49.

Coincidence?

He wondered if the object, whatever the hell it was, might be giving off some kind of magnetic impulse. Perhaps that was what had caused the radio to blow out too.

The object disappeared behind a range of hills.

David cast anxious glances to the spot in the sky where it had been but could not see it.

'It's gone,' said Julie.

David didn't answer. He slowed down slightly, scanning the heavens for the yellow shape, almost relieved when he didn't see it.

'If it was a plane, perhaps it landed somewhere,' Caroline offered.

'Perhaps,' David replied. 'Though I don't know where.

There isn't any ground flat enough to bring a plane down for miles.'

'Perhaps there's been a crash, Dad,' said Dean, almost excitedly.

David didn't answer, he was still scanning the sky for the yellow egg, thinking that it would appear again at any second.

It didn't.

He exhaled, relaxing his grip on the wheel. A smile even touched his lips.

'Well, everybody, it looks like our excitement is over for the night,' he said, wondering if the others would realize how happy he was to be saying those words. 'We should be home soon.'

Dean clambered up on the back seat and peered out of the rear window, gazing into the sky.

David slowed down and swung the Volvo right at Amber Hill, the vehicle bumping for a moment as it travelled over the rough surface. David rubbed his eyes and smiled across at Caroline who had settled down in her seat once more, eyes closed. In the rear view mirror he glimpsed Dean still up on his knees, looking out of the back window. Michelle was staring out of her window, watching as the countryside sped past in the darkness, her head bumping against the glass as the car rolled on.

Hills rose up on either side of the road, dark shapes which seemed to grow from the night itself. To the left was the black mass of Cocking Tor, masked for the most part by thick expanses of forestry commission planting. The Volvo seemed to be the only thing moving in the blackness. David fiddled with the frequency knob on the radio, trying to find something. News, music, anything. The dial, however, was still hot. The hands of both his own watch and the car clock were still immobile at 10.49. David shuddered involuntarily, then smiled to himself, trying to dismiss what had just happened as nothing more than 'a little adventure'. He shook his head. How empty and unconvincing those words sounded. Who was he trying to fool? He had been frightened, as, he imagined, anyone in a similar position would have been. The sight of the unidentified . . .

He allowed the thought to trail off.

Of the what? The UFO? The flying saucer? Had that been what they had witnessed? David reached for the packet of cigarettes on the parcel shelf and lit one. He took a long drag on it and blew out a stream of smoke.

'Dad.'

The voice came from behind him.

'Yes, Dean,' he said.

'That light we saw.'

David smiled again, wondering what questions his inquisitive son was going to put to him. The boy was, as is the way of ten-year-olds, curious about everything he encountered, never resting until he had an answer. David wondered what he was going to tell his son if the boy asked about the possibility that they'd seen a UFO. Should he tell him the possible truth and risk frightening the child? He waited for the question.

'What about the light?' he said.

'It's behind us.'

David spun round in his seat, almost sending the car off the road as he tried to see out of the back window.

It was there all right. Twice as large as before, much closer to them, its yellowish glow now so profound it was white in places.

'Oh God,' murmured David.

'Dad, it's getting closer,' Dean informed him. 'I think it's following us.'

David stubbed out his cigarette in the nearby ashtray and, glancing quickly in his rear-view mirror, put his foot down. The Volvo shot forward, tyres skidding on the uneven surface. This time, framed in the mirror, there was no mistaking its shape or size.

It was, indeed, egg-shaped, and David guessed its size to be somewhere close to thirty feet across. It was moving steadily above the road, skirting tree tops in places and he reckoned that it could be less than twenty feet from the ground. How close it was to the car he didn't like to imagine. Another anxious glance in his rear-view mirror told him that the craft was perhaps 500 yards behind, but the distance was shortening by the second. The brilliant glowing shape cast no illumination. The road and hills remained in darkness despite the searing luminosity radiated by the craft.

But as it sped nearer to the Volvo, the interior of the car was filled with light and David waited for the moment when it would pass over them.

That moment did not come. The craft seemed to slow down to a cruising speed, content to stay behind the fleeing Volvo. Almost as if it were playing a game, it had shown David that if it needed to it could catch up with the car effortlessly, but it remained content to hang back a few hundred yards. Waiting.

He stared at it in the rear-view mirror until his eyes hurt.

Caroline, and indeed all the family, were looking out of the back window at it. No one spoke. Fear was etched on their faces. Only Dean watched the object with something akin to fascination.

'What is it?' Caroline said, not taking her eyes from the pursuing craft. 'Dave, what . . .'

He cut her short. 'I don't know,' he barked, flooring the accelerator, struggling to retain control of the Volvo. The car hit a ramp of earth and, for interminable seconds, left the road. It hung in the air for a moment then slammed down on to the tarmac again, skidding as it did so. Michelle screamed. Julie pulled her close, shielding the child's eyes from the blinding light behind. She herself had her eyes closed, hardly daring to look round.

David felt terribly afraid. Not for himself but for his family. What if he should swerve off the road, crash the car? What if the craft should attempt to halt them? What if . . . He tried to push the questions to one side, frightened at the answers he might find. He gripped the steering wheel until his knuckles turned white and watched as the needle on the petrol gauge began to tilt lazily up and down once more from full to empty. And it was then that the thought struck David.

With the gauge behaving so erratically, he had no way of knowing whether or not he had enough fuel to get them home. What if the car ran out of petrol now?

He gritted his teeth and refused to think such thoughts. Along that path lay true insanity.

Another look in the mirror showed him that the object was keeping a measured distance from the car, almost as if it were waiting for the right moment. David dared not drive any faster, especially with the thought that he might run out of fuel lurking in the back of his mind. He tried desperately to think of something to do, anything to escape the pursuing craft. The road was so desolate, so isolated, he doubted if above a dozen vehicles used it in one day let alone at this time of night.

He heard a low buzzing in his ears, a sound which grew in volume until it seemed to fill the entire car. Or was it just his head? Was the sound inside his head?

There were lights ahead.

He glanced down at the fuel gauge once more to see the needle nudge red. Please God let them reach the lights ahead. It looked like a house. The Volvo's powerful headlights cut through the blackness and pinpointed some wooden fencing. Nailed to it was a sign:

ANIMALS. GO SLOW.

A farm, he thought. They were almost there, too.

David swung the car into the yard, narrowly missing a metal post. He was thankful that the gate was open for, had it not been, he would have ploughed straight through it. The lights in the farmhouse beckoned.

Behind them, the craft soared upwards and veered sharply off to the left, disappearing behind some tall trees.

David stepped on the brake and brought the Volvo skidding to a halt in front of the farmhouse. He jumped out of the car, noticing that the front door of the building was already opening. Light spilled into the yard as two men emerged from inside. One was carrying a shotgun.

'Can I use your phone?' David blurted, praying that they had one. He could not see the men's faces, framed, as they were, by the light from behind them.

'I have to call the police,' he said. 'Please.'

'What the bloody hell's going on?' said the first, and older of the men.

'Just let me use your phone, please,' David repeated.

The older man nodded and looked across the yard at the car. He watched as David ran back and ushered his family from the vehicle towards the farmhouse, all the time glancing up at the sky as if expecting the craft to appear again at any moment. He had seen it fly off, rising high into the air before plummeting down again. Where it was now he could only guess. Now the most important thing was to see his family safely inside the house and then get help.

The older man, Alan Hughes, put down his shotgun when he saw the looks of fear on the faces of the women and children approaching from the car. He and the other man, who, it transpired, was his twenty-three year old son, Gary, helped the family inside, then shut the front door.

A woman had appeared at the top of the stairs in her dressing gown. A large rotund woman who David took to be the farmer's wife.

'What's happening, Alan?' she asked her husband.

'These people are in trouble,' the farmer said.

The woman, Vera, came bustling down the stairs and led the family into the sitting room, ensuring that they were comfortably seated. She studied the array of white faces before her and pulled the cord of her dressing gown tighter around her.

'You lot look as if you've seen a ghost,' the farmer said.

David swallowed hard. 'Can I use your phone, please?' he said, breathlessly.

'Look,' Alan Hughes began. 'You can use the phone, but before you go dragging the police out here I'd like to know what's going on.'

David sat down on the corner of the nearby sofa. 'We were chased by something,' he said. 'I know it sounds ridiculous but it was like a . . .' He hesitated to use the word. 'Like a spaceship of some kind.'

Gary Hughes chuckled. 'A flying saucer,' he grinned.

'I don't know what it was,' David snapped. 'There was a light, a huge light. It followed us.'

There was silence for a moment.

'So where is it now?' the farmer wanted to know.

'It flew off in that direction,' said David, motioning towards the last spot he'd seen the craft. 'It looked as if it might have landed.'

'That's Blakelow Hill over that way,' Hughes told them. 'There's a disused quarry just beyond that.'

'Perhaps it landed there,' said Caroline, sweeping the hair from her face and pulling Michelle closer to her. The child was obviously still terrified and Caroline could feel her quivering.

'You call the police if you like,' said Hughes, heading for the hallway and returning with his coat and wellington boots. 'I'm going to have a look for this bloody thing, whatever it is.'

'Why don't you just let the police handle it?' David said, getting to his feet.

'Because if there's someone on my land, I want to know about it,' Hughes said, picking up the shotgun.

'Perhaps there's some little green men,' said Gary, smiling again.

David glared angrily at him. 'What do you think this is, a joke?' he rasped. 'Do you think I'd have come to this place if there hadn't been a good reason?'

Gary coloured and lowered his gaze.

'I'm going with you,' said David, moving towards Hughes.

'I thought you wanted to call the police,' the farmer reminded him.

'If there is something out there then I want another witness.'

Hughes nodded. 'Gary, you stay here, we'll go and have a look over in the quarry and across the fields around Blakelow Hill, see what we can find.'

'I'll make a pot of tea,' Vera added. 'You look as if you need it.' She took one more look at the two women and the children huddled before her and scuttled off to the kitchen.

David paused at the front door then peered into the sitting room once more. 'Caroline,' he began. 'Call the police, as soon as we leave. Tell them what happened. Just get them out here.'

She nodded.

Hughes opened the door and the two of them walked out into the darkness. David felt a chill run up his spine despite the warmth of the night.

The warmth. It was the first time he had noticed it. The air itself seemed unusually hot, as if they were sitting close to an electric fire, a kind of prickly heat which made him feel uncomfortable. In all the excitement he had not noticed it until now. The air was almost crackling and he felt the hairs at the back of his neck rise. Could it, he wondered, be static electricity of some kind? Hughes, too, noticed it for he paused in the yard and wiped a hand across his face. The barrels of the shotgun felt warm to the touch.

The farmer paused for a moment, looking behind him.

'What's wrong?' asked David.

'The dog,' Hughes told him. 'I don't know where the dog's gone. The daft bugger'd been howling for about five minutes before you arrived, then he just cleared off into the kitchen.' Hughes smiled. 'Perhaps he knows something we don't, eh?'

David tried to smile but couldn't.

They set off towards the place where the craft had disappeared, and, despite the increased heat of the night, David felt his skin tightening. He glanced anxiously at the sky and walked on.

Two

'Here, take this,' said Hughes, handing David the torch which he had picked up from the table inside the hallway.

The farmer paused as David flicked it on, the broad beam cutting through the darkness. Then, side by side, they moved off across the yard towards the low fence which marked the perimeter of the nearest field.

Hughes steadied the shotgun in his grasp as he scrambled over, glancing at his companion to see that David was looking alternately at the fields and trees ahead and also at the sky. The farmer brushed a hand across his forehead, perspiration forming in salty beads there. He wondered why it felt so damned hot. The hair was prickling on his scalp and hands, standing up like the hackles of a frightened dog.

The two men moved slowly across the field. The only source of light came from the torch which David held. As he scanned the area ahead of him he could see no sign of the yellowish glow.

'What did you say was in this direction?' he asked Hughes.

'Some woods,' the farmer told him. 'Hills and an old quarry.'

David nodded.

Perhaps the blasted thing had gone to ground in the quarry. That was why they could see no sign of the powerful light. He swallowed hard, finding that his throat was dry, as if he'd gone without a drink for many days. His skin, particularly on his face and neck, felt as if it had been rubbed with a scourer and he found it difficult to breathe in the growing heat. To David there seemed no doubt about it.

It was getting hotter.

To him it was like walking through a greenhouse where someone was slowly raising the temperature. However, there was none of the humidity in the air which he associated with greenhouses. The light from the torch wavered as he wiped his forehead.

'That's bloody curious,' Hughes said, looking round.

David glanced at him. 'What's wrong?' he wanted to know.

The farmer pointed ahead of them, quickening his pace. David followed, bringing the light to bear once more. The beam picked out a dark shape on the ground about twenty feet from them. He gritted his teeth.

It was a cow.

The animal lay quite still and Hughes approached it cautiously, as if he were walking towards someone who he didn't want to wake up. It certainly didn't look as if he was going to disturb the cow, which showed no sign of movement as the men drew closer.

It was lying on its side, head tilted backwards and Hughes frowned as he reached it.

'Give me the torch,' he said, holding out a hand to take it, bending over the motionless carcass, shining the powerful beam over the body.

He rested one hand on its side and felt the barely perceptible rise and fall.

'Is it dead?' asked David, looking first at the animal and then anxiously at the sky ahead of them.

'No,' Hughes told him, a puzzled expression on his face.

Ten feet further on lay another cow, in a similar position. Close by there were more. None of them moved, but as the farmer checked each one he found that they were all alive. There were no visible signs of injury to any of the beasts, despite the fact that each one had gone down as if pole-axed. Hughes shone the torch across the ground around them and saw that there had been little or no movement from the animals prior to them collapsing. The grass and earth were not trampled or pawed. It seemed that they had, en masse, been felled by something which had put them all into a deep sleep.

Hughes straightened up, finding now that *his* breath caught in his throat.

'It's like they've been drugged,' he observed, prodding one of the fallen animals.

David could only shrug his shoulders. He took the torch from the farmer, noticing that the light from it seemed somehow dimmer. He shook it, wondering if the batteries were running down. It felt clammy against his sweaty palm and he wiped his hand on his trouser leg. The torch flickered once, then glowed again but with less brilliance. David walked on, glancing at Hughes who was looking around him at yet more fallen cows. The field seemed to be full of them.

'Listen,' said David, stopping.

Hughes slowed his pace. 'What is it?' he asked.

'Can't you hear it?'

The farmer didn't know what he was supposed to be hearing.

David looked almost imploringly at him, wondering why he could not hear the noise, wondering if he was imagining it. Were his ears now beginning to play tricks on him too?

Hughes stood still and listened.

David looked at him, watching for some reaction, but the older man seemed to hear nothing. What was wrong with him, was he deaf? Why couldn't he hear it?

David thought back to the happenings inside the car earlier. Had he imagined it then? Had the sound merely been inside his own head, as he'd suspected? He sucked in a worried breath and looked, once more, for any sign of recognition from Hughes.

'Can't you hear anything?' he asked the farmer.

Hughes suddenly spun round, ears cocked in the direction of the noise. He swallowed hard then nodded. 'I hear it,' he said, quietly.

David felt something akin to relief. At least he now knew that his senses weren't playing tricks on him.

The humming sound seemed to grow louder, reaching a

certain pitch then levelling off. It was like the monotonous single note of a phone which has been left off the hook, but it was much louder.

Both men stood still for long minutes, trying to pinpoint the direction from which the sound was coming. It was Hughes who realized that it lay ahead of them.

The looming black outline of a hill masked the terrain beyond and David felt the ground beginning to slope upwards as they pushed on. Trees grew in erratic clumps on the slope, gradually drawing together as the hill gently peaked. He shone the torch before them, its dull beam picking out the pines and cedars which stood like sentinels, silently overlooking the land.

The buzzing grew louder as they approached the top of the hill.

David felt like stopping where he was and he visibly slowed his pace. Even Hughes, with the added security of the shotgun, was treading more carefully. The incident with the cows had unsettled him, but now this! But, overcoming their mutual apprehension, both men pushed on until they were within yards of the crest.

It was then that the torch went out.

David shook it violently. He flicked the on/off switch, simultaneously angry and worried when nothing happened.

The buzzing was almost deafening by now and he looked over to see Hughes raise a hand to cover one ear.

They reached the top of the hill and looked down.

The ground below them was thickly wooded, rising here and there in a series of gentle undulations.

Nestled amidst a clump of pines, its shape now clearly defined, was the yellow, egg-shaped object.

'Christ,' muttered Hughes. 'What *is* that?'

David didn't answer him. Firstly because he had no answer to offer and secondly because he barely heard the words. His attention was riveted on the object, as he tried to make out exactly what it looked like. It was, as he had originally thought, oval, although it did seem to be elongated at either side making it appear more like a cylinder. In fact, as he squinted through the powerful light he saw that the thing was almost cigar-shaped. It radiated so much luminosity that its shape, at first glance, was not easily discernible. Even as he looked, it seemed to contract, to change shape once more into that all too familiar ovoid which he had first seen. David also noticed that the object was not, as he had first thought, suspended *amongst* the trees, it was hovering slightly above them.

'What is it?' Hughes repeated once more.

David felt the hair rise on the back of his neck and, as before, it felt as though his skin were being rubbed by something harsh and coarse. He shielded his eyes, unable to tear his gaze from the object before him. He squinted. It felt as if someone had thrown sand into his face, the grains rubbing back and forth across his eyes.

The buzzing stopped abruptly but neither man seemed to notice it.

Hughes took a step backwards as the craft began to rise, foot by foot, gradually leaving the trees until it hovered about thirty or forty feet above them. It ascended in a perfect vertical, moving with infinite slowness, almost as if it wanted the men to get a good look at it. Then it seemed merely to hang in the air.

David also moved backwards, gripping the trunk of a tree as he watched the yellow shape rise. For what seemed like an eternity, it remained motionless in the sky, held there by a power which neither man could hope to comprehend, then noiselessly it sped upward at terrific speed leaving nothing but an afterburn on the retinas of the reluctant spectators. Both men blinked, looking away as the intensity of the light grew too much for them and for brief seconds David thought he had been blinded.

When he looked back, the light had become darkness.

Of the craft there was no sign.

Hughes was still staring ahead at the spot in the sky where it had been. His brow was heavily furrowed, his mouth open. The shotgun rested in the crook of his arm. David doubted if he would have been able to use it even if the need had arisen.

The farmer could only stand motionless, his ears still full of the buzzing, his eyes still stinging from the intensity of the light.

'Jesus Christ,' he finally managed to murmur. 'What *was* that?'

David exhaled deeply and loosed his grip on the tree trunk, swaying uncertainly for a moment. 'We'd better get back to the house,' he said, turning and starting down the hill.

The torch suddenly burst into life and the men found that the beam had regained its customary brilliance. They hurried back down the hill towards the field, anxious to reach the house once more. Hughes now also found himself looking up at the night sky wondering if the mysterious craft was going to reappear.

'I didn't believe you,' he said, almost apologetically. 'When you first told me. I didn't believe you, I . . .'

David cut him short. 'It doesn't matter now,' he said.

At least now they had an independent witness, someone from outside the family who had seen the object at first hand. Should the police doubt them, Hughes would be there to back them up.

David found it easier to get his breath now, and the night air felt cool on his hot skin. Only his eyes still hurt; the skin on his face and neck merely itched now. He resisted the urge to scratch it.

The two men found that they were practically running in their eagerness to reach the house, as if the craft were going to reappear at any minute.

Something moved ahead of them and they froze, listening to the sounds of movement in the darkness.

David shone the torch ahead of him, the beam picking out several shapes.

'What the hell is going on around here?' said Hughes exasperatedly.

Every single cow in the field was on its feet and moving around.

Three

As David and Hughes drew nearer the farmhouse they both caught sight of the police car parked in the yard.

'Thank God,' murmured David, as the two of them climbed the fence from the field.

Hughes didn't speak, the incident with the cows had shaken him badly and he moved now in a kind of daze. David knew how he felt. The sight of the police car reassured him to some degree, but even so he could not resist a wary look skywards as he and Hughes reached the front door of the farmhouse.

It was opened by Vera Hughes who ushered them into the sitting room.

David saw two policemen in the room, both constables. One was in his forties with greying hair and a thick moustache, the other a younger man with a small scar on his left cheek.

'How long ago did you arrive?' David said, looking at the two uniformed men.

'About five minutes ago, sir,' said the elder one.

'Did you see it, then?' asked David.

The policeman scratched his head. 'See what, sir?'

'The light. The . . . craft?'

David crossed to the sofa where the rest of his family were huddled.

Vera and Gary Hughes stood beside the chair into which Hughes himself had slumped, a look of bewilderment on his face.

'I'm constable Vincent,' said the younger man. 'This,' he indicated his companion, 'is constable Bryant. Your wife has told us about what you think you saw.'

He produced a note book which David could see had been scribbled in.

'I don't *think* I saw anything,' David snapped. 'I *did* see something. Have you questioned my family?'

Vincent nodded.

'It is late, sir, you're probably tired. The eyes can play all kinds of tricks,' Bryant added.

'We saw something,' David said through gritted teeth. 'A light in the sky. A UFO. Call it what you like. I was as sceptical as you to begin with but I'm telling you, we all saw it.'

'I saw it too.'

The voice came from Hughes and all eyes turned in his direction.

'What exactly did you see, Alan?' asked Bryant, flipping open his own note book and perching on the arm of the chair.

Hughes explained about the cows, how they had seemed to be unconscious, how he and David had heard the buzzing sound, seen the yellow ovoid and watched it disappear and then how the cows had miraculously recovered with the departure of the thing.

Bryant wrote it all dutifully down. 'Is there any physical evidence of any kind to support the sighting?' he asked.

David looked blank for a moment, glancing at Caroline.

'We saw it, constable,' she said. 'What more do you need to know?'

'If you'll excuse me saying this, Mrs Ellis, if we go back to the station and file a report like this, they'll think *we're* nuts.'

'Are you saying *we're* bloody crazy, then?' David demanded.

Bryant shook his head. 'No one said anything like that, sir. It's just that I'd like some physical proof.' He closed his note book and sighed. 'We don't know any more about flying saucers than you do. I don't know what you saw.'

David opened his mouth to speak but Bryant held up a hand to silence him. 'Your wife said that something happened to your car,' said the policeman.

'Yes,' David told him. 'The dashboard and wheel got hot. The radio seemed to overload.'

Bryant got to his feet. 'There might be some kind of physical sign on the vehicle, we'd better check that.' He nodded towards Vincent who followed him towards the front door.

David kissed Caroline and the children briefly then wandered out after the policemen.

'Dad,' said Gary Hughes, looking at his father who was still staring blankly ahead of him. 'What happened out there?'

'We saw this light,' Hughes told him. 'I've never seen anything like it before. I thought it was going to blind me. Then it just disappeared.' He clicked his fingers. 'As if it had never been there in the first place.'

'Where was it, Dad?' asked Gary. 'Whereabouts on the farm?'

'In the woods past Blakelow Hill.'

There was a long silence.

'I don't know what it did to the cows,' Hughes continued.

'Were any of them hurt?' Gary asked.

Hughes shook his head. 'It just looked like they were stunned. Knocked out. I thought they were dead at first.'

'How long ago did you see the . . .' Caroline struggled to find the word. 'This . . . light.'

Hughes shrugged. 'About ten or fifteen minutes after we left the house.'

The farmer suddenly seemed to come out of his trance-like state and his features became more animated. 'Why?' he asked.

'Because all the lights in the house went out,' Vera told him. 'Even the fridge switched itself off.'

'The lights were off for about seven or eight minutes,' said Gary. 'That would have been about the same time *you* saw the light.'

Constable Vincent walked slowly around the Volvo, shining his torch over the chassis of the vehicle. Bryant and David stood to one side watching him, both of them scanning the car for any signs of damage or an indication of what had happened. David swallowed hard, willing the constable to find something. He continued his careful inspection.

'You say the steering wheel and dashboard got hot when this light appeared?' asked Bryant.

David nodded. 'The radio blew out too. I couldn't get anything out of it.'

'Could it be something wrong in the engine?' Bryant asked. 'Perhaps that, or your heating system is not working properly.'

'For God's sake, constable,' said David, in frustration. 'For one thing, it's July, I'd hardly need the heater on, would I? And for another, the car was only serviced two weeks ago. It was nothing to do with anything inside the car. It was the heat from the spacecraft that was doing it. The needle on the petrol gauge was playing up as well. It was as if we were riding under some huge sun-ray lamp.'

'Could I just have a look inside the car, sir?' Bryant asked, holding out his hand to take the keys.

David handed them over somewhat reluctantly.

'No sign of any damage on the outside,' said constable Vincent.

'Check under the bonnet will you, John?' asked Bryant, unlocking the driver's door and clambering inside.

David stood beside him and showed him where the dashboard had grown hot. Bryant leant across and felt around beneath the parcel shelf.

'You may have got a loose wire somewhere,' he said. 'Something could have shorted out and caused the heat.'

David lowered his head, trying to keep a tight rein on his rapidly fraying temper. 'There's nothing wrong with the car itself,' he breathed.

'Nothing wrong with the engine, anyway,' Vincent confirmed, slamming the bonnet down.

Bryant started the engine and pressed down on the accelerator, revving the engine until he saw the needle on the petrol gauge rise to about half-full. He eased the pressure then finally switched off the ignition.

'Everything seems to be fine, sir,' he said.

'So, is that it, then?' said David, angrily. 'Investigation over.'

Bryant clambered out of the Volvo, shut the door and handed the keys back to David. 'Unless you want to make a formal report, sir, there's not much we can do.'

David banged the bonnet of the car angrily. 'My whole family saw it,' he rasped. 'Mr Hughes saw it. What more evidence do you need?'

'Eye-witness reports in cases like this aren't usually very reliable,' Bryant told him almost apologetically.

'Do you think we all had the same . . . vision, suffered from the same illusions at exactly the same time?'

'No, sir, but I need physical proof and, even if I had that, it's doubtful that anything would come of this report.'

'Why not, for God's sake?'

'Because that sort of thing isn't our business. We're used to dealing with burglars, shop-lifting and the odd lost cat, not flying saucers. No one's interested.'

'My family and I could have been in danger, isn't that of any interest to anyone?' David snapped angrily.

'The report would probably be filed away somewhere,' Bryant told him. 'No one would ever see it.'

'Jesus,' muttered David.

'But it's up to you, sir,' Bryant continued. 'If you want to make the report then fair enough, but . . .'

David cut him short. 'But don't expect anyone to believe it,' he said, acidly.

'Would *you* have believed it if someone else had told *you*?' the older policeman wanted to know.

There was a long silence. David looked up at the sky, now clear, dotted with the occasional star, brushed over by smoke-like wisps of cloud. He almost willed the yellow shape to appear above them, so that he could shout to them, 'There it is, do you still disbelieve me?' But the sky remained clear and David chuckled softly to himself. He looked at the two policemen.

'I suppose you're right,' he said. 'I wouldn't have believed anyone else if they'd told me. I don't blame you for being sceptical.'

'It's not that we don't believe you, Mr Ellis,' said Bryant almost apologetically. 'But, well, there is no physical evidence. You know how it is. I realize that you saw something. You don't strike me as the sort of man who'd drag us out here in the middle of the night on a wild goose chase, and I've known old Alan in there,' he hooked a thumb behind him in the direction of the farmhouse, 'for years. He hasn't got that sort of imagination. I don't doubt that you saw something, but I'm afraid there's nothing more we can do to help you.'

David nodded.

'We can drive you back home if you like,' offered Vincent. 'Just in case. Have you got far to go?'

David shook his head. 'Matlock,' he told them. 'Well, just outside the town actually. About fourteen or fifteen miles from here.'

'We came from Matlock, we can wait for you,' Vincent persisted.

'No thanks, we'll be fine now,' David told him.

The two policemen nodded and retreated to their waiting Panda car. David watched as Vincent started the engine and guided the vehicle through the gate, back on to the pockmarked road. The tail lights gradually disappeared into the darkness.

David turned and headed back into the farmhouse.

He found Caroline waiting for him in the hall. 'Well?' she said. 'What are they going to do?'

He repeated his conservation with the policemen, watching as her expression changed from one of bewilderment to one of anger.

'So they're not going to do anything?' she said when he'd finished speaking.

David shook his head. He pulled Caroline closer and kissed her on the forehead. 'I can understand their point of view,' said David. 'Even if they *had* found proof what could they have done?'

They stood in silence for a moment then made their way back into the sitting room where David repeated the story once more, for the others. He waited for their reactions but there were none. Caroline and Julie both looked tired and the children, Dean in particular, were on the verge of falling asleep. David looked down at his watch and then across at the clock on the Hughes' mantelpiece. It was just after midnight and David was both surprised and relieved to see that his own watch had started again. He pulled out the winder and adjusted the hands to the correct time: 12.03 a.m. Time seemed to have lost its meaning. It seemed like an eternity since the bizarre events began over ninety minutes earlier.

'We'd better be going, we've taken up enough of your time,' said David, ushering the rest of his family to their feet and out into the hall. He shook hands with Hughes. 'Thank you for your help.'

'Did the police say whether or not anyone else had reported seeing that light?' asked the farmer.

'If anyone did see it, either they didn't report it or the police aren't saying anything,' said David.

'Will you be all right driving home?' asked Vera, watching as the family bundled into the waiting Volvo.

David nodded but, despite himself, he couldn't resist a quick glance at the sky. He shook hands with Hughes once more then hurried over to the car, got in and started the engine. Within minutes they had left the farm behind.

He drove slowly along the bumpy road, squinting through the darkness for signs which would direct him back on to a main road and, more importantly, home. David was anxious to be inside and hidden from any other possible incidents. He looked down and saw that the clock on the dashboard was also working again. Behind him, Dean had stretched out across Julie's lap and was fast asleep. Julie herself was dozing, her head resting on Michelle's who was also asleep.

David looked across at Caroline who smiled softly at him,

reaching across to squeeze his hand.

'Are you OK?' he whispered.

She nodded and smiled again, closing her eyes.

'Soon be home,' he told her, and decided to see if he could get anything coherent out of the radio. As he turned the dial it buzzed into life and there was a low hiss of static as he tried to find a music channel. The stations flickered through. A discussion programme. The news. He listened to it for a moment then continued turning until he found some music. David smiled to himself for a moment until he heard the song. David Bowie was singing 'Space Oddity'.

He switched the radio off.

Four

David stifled a yawn as he turned the car onto the main road which would take them home. A sign opposite, pin-pointed in the Volvo's headlamps, announced the fact that Matlock lay a mere ten miles away now. Living as they did about five minutes drive out of the town, David realized that it should take them but a short time to reach their destination.

The rest of the family were asleep, he noted. Just as well, perhaps, he thought. After what they'd been through earlier it was little wonder they were exhausted. David himself felt tired. Drained both emotionally and physically. As he drove he touched his face and winced. That all too familiar prickly feeling began to afflict his skin but he resisted the urge to scratch it, gripping the steering wheel tighter instead.

A car passed them going in the other direction and David felt almost relieved that there were still other vehicles on the road. But as soon as the other car had passed, he once more felt exposed on that wide road. Hedges rose on either side of it, blending in with the darkness, and only the glint of the cats' eyes in the road offered any illumination other than the Volvo's headlights.

Caroline stirred slightly beside him, brushing some strands of hair from her face. She looked flushed and David noticed the beads of perspiration on her forehead. Indeed, he began to feel his shirt sticking to his own back. It seemed to be getting warmer inside the car. He leant forward and fiddled with the heater but it was turned off, as he had suspected. He reached across to wind down the window.

'Damn,' he rasped, pulling his hand quickly away from the metal handle.

The metal felt hot, and now he noticed the steering wheel too was getting warm. The windscreen began to mist over and David hurriedly wiped it clean, his breath now coming in gasps. Was the engine overheating, he wondered? He checked the temperature. It was fixed at normal. Besides, he reasoned, if the engine were becoming too hot it wouldn't cause the metalwork inside the car to heat up.

Caroline woke up and groaned. 'David, it's so hot in here.'

'I know,' he replied, adjusting the heater to blow cold. But when he switched it on, a blast of heat poured forth and he shut the valve off quickly.

'What the hell is happening?' he muttered, angrily.

There was a lay-by about fifty yards ahead and David pulled into it. He switched off the engine and sat motionless behind the wheel for a moment, panting heavily. It seemed as if all the air had been sucked out of the car. The interior of the vehicle smelt fusty, dry.

'What's wrong, Dad?' asked Dean sleepily from the back seat. Then he too began to wriggle as the heat wrapped itself around him like a clammy glove.

'I'm going to check under the bonnet,' David announced reaching for the door handle. That too was hot, but he disguised his pain on touching the hot metal and clambered out of the Volvo, slamming the door behind him. The night air carried an unnatural warmth and David shuddered in spite of the heat as he raised the bonnet of the car. He pulled the torch from his jacket pocket and shone it around over the engine. The heat hit him like a palpable blast and he recoiled. Even the water inside the windscreen cleaner was bubbling. David stepped back and shook his head, looking down into the engine.

He heard the humming sound almost at once.

His heart hammering against his ribs he spun round.

The road both behind and ahead was empty and yet David could hear the humming sound and it seemed to be getting louder. He swallowed hard and looked up.

High above him, suspended in the air, was the yellow, glowing craft.

'No,' murmured David.

He slammed the bonnet of the car down, feeling the heat in the metal, and scuttled back to the driver's side door where he clambered in and started the engine, praying that the heat had not made it seize up.

'Did you find out what was wrong?' Caroline asked him.

David shook his head, turning the key frenziedly until the engine roared. He stuck the car in gear and stepped on the accelerator. It shot forward, causing Julie and the children on the back seat to lurch over momentarily. David floored the accelerator, not daring to look out of the window for fear of spotting the glowing yellow object which he had seen.

He felt like shouting aloud that it was back. The fear and frustration and . . . anger? Yes, it was anger. It welled up inside him until it seemed he could not restrain himself any longer. Eyes locked on the road ahead, hands gripping the hot steering wheel, he drove like a madman.

The luminous craft appeared before them like something from a shared nightmare and Caroline bit her fist to stifle a scream.

David guessed that it must be about two or three hundred feet above them, travelling in the same direction but, no mistake, it was dropping lower in the sky all the time.

'David, what are we going to do?' asked Caroline, but he had no answer for her. He glanced briefly at the ovoid, then forced himself to stare at the road once again.

There was an intersection approaching and David knew he would be forced to slow down, a thought which sent a renewed shiver of dread through him. By this time, the glowing yellow egg had dropped to less than fifty feet, and David noticed it was heading towards the car, not away from it.

The glow grew brighter as the craft approached at seemingly impossible speed. David looked up at it, mouth agape.

He didn't even see the car which cut across his path.

There was a screech of brakes and the squeal of spinning tyres as the two vehicles avoided one another by a matter of feet. The other car kept going, the driver banging his hooter angrily. David wondered if *he* had seen the craft too. If he had, he certainly didn't hang around for a better look. In a matter of seconds, the car had disappeared and the road was empty once more. David looked up at the yellow shape which was now speeding towards the Volvo with the speed of a bullet, the glow becoming white, so intense was it.

In the back seat, Julie and the children huddled down, shielding their faces from the oncoming craft. Caroline too turned away, whimpering in fear. Only David watched, with the same terrified fascination with which a mouse watches a snake, as the blazing orb drew nearer. It seemed just a matter of seconds before it struck the car. The buzzing sound which he had heard in the woods at Alan Hughes' farm began to fill his ears until it was the only thing he could hear. He opened his mouth to scream but no sound would come.

The craft hurtled ever lower, gathering speed all the time.

Was this to be the end, David wondered? Were they all going to be killed now? He gripped the hot wheel tighter and whispered a prayer.

The buzzing inside his head grew to deafening proportions. At last David found the breath for a yell of rage and fear.

The craft swept over the car, missing it by a matter of feet and David saw the glowing ovoid in his rear view mirror. It swung in an incredibly narrow arc and came hurtling after the car once more, and for the first time David noticed that apart from the buzzing it made no sound at all. Watching the object carefully, he pressed down harder on the accelerator in a vain attempt to outrun the pursuing craft.

'David, it's trying to kill us,' said Caroline, still crouched in her seat, hands covering her head.

He didn't speak, but instead swung the car to the right onto the B-road which would bring them to their destination. He glanced down again at the temperature gauge and saw that the needle was touching red. The engine was overheating.

'Please God let us make it,' said David aloud, as the shape swooped past them once more. But this time it rose. Higher and higher until the blazing yellow luminosity became just a pin-prick amongst the other stars. For interminable seconds it hovered, then, as David continued to wrestle the car under control, the craft began a descent which seemed certain to condemn it to a crash landing. Like a bullet it sped down from high above, aimed directly at the car.

'Come on, come on,' David shouted, as the car began to falter, the engine now hissing steam. But there, a few hundred yards ahead of them, was the pub which they owned and ran.

Home.

Please God that they reach it in time.

The craft was still speeding towards them like ball lightning, a steadily growing single point of brilliance which drew closer with each passing second.

The pub was now less than a hundred yards away and David found himself banging the steering wheel in rage and frustration as he saw the craft getting closer.

And then they were there.

The car screeched to a halt in the yard and David threw open the driver's door, digging his hand into his coat pocket, fumbling to pull out his door keys. He searched the key ring for the right one, found it and twisted it in the lock, then he ran back to the car and ushered the rest of his family out. They ran for the open front door, the craft now apparently filling the sky with its dazzling glow.

David looked behind him as he ran and saw that the yellow ovoid showed no sign of slowing down.

He pushed the others inside before him, then slammed the door, hastily sliding the bolts, not thinking how little use they would be if the craft *did* hit the building. Something which, from the speed it was travelling, seemed unavoidable. David rushed to the nearest window and watched as it rocketed towards the pub. The glow was blinding now, yellow, then white until he was forced to cover his eyes.

He dropped to the floor and waited for the impact.

It didn't come.

The craft stopped dead, hanging in the air as if some

gigantic invisible hand had prevented it hitting the pub. It hovered soundlessly outside, about fifty or sixty feet from the ground.

David slowly raised himself up and peered through the glass at it.

'Phone the police,' he said to Caroline, who hurried through into the sitting room to do so. She switched on the light and grabbed the phone, her shaking finger dialling three nines.

There was a series of loud bangs and Michelle screamed.

The sounds came from below them. From the cellar.

'David,' called Caroline. 'The phone's dead.'

He realized that the sounds from below must have been a fuse blowing. Could the craft somehow be draining power from the pub, David wondered? He watched it as it seemed to glow even brighter.

Caroline returned from the sitting room. 'What are we going to do?' she blurted, hugging Michelle close to her.

David didn't answer, his eyes were still riveted to the glowing orb outside.

'David.' She practically screamed at him.

The buzzing sound filled his ears, his eyes felt as if they were being burned from the sockets and, finally, he was forced to turn away. The glow from the object filled the pub and the family could only crouch helplessly, bathed in the unearthly yellowish-white light. Both the children now were crying and Caroline and Julie did their best to comfort them. David crawled across the floor to join them.

The room was suddenly plunged into darkness.

For a moment David thought that his sight had gone, but as he rubbed his eyes he saw his wife, her cheeks stained with tears, a look of pure terror on her face.

He chanced a look towards the window.

Of the craft there was no sign.

He pressed his face to the glass, surprised to find it slightly warm. He scanned the heavens.

Nothing.

Caroline looked up to see him heading for the front door.

'Don't go out there, David,' she said, a note of urgency in her voice.

But his hand was already turning the key.

'David. No.'

He opened the door and stepped cautiously outside, the breath held in his throat.

The sky was clear. The craft was gone, having vanished as swiftly as it had come.

Something large bolted past him from inside the pub.

David jumped, his heart skipping a beat until he realized what it was.

King, largest of the family's two Alsatians, was standing in the yard barking loudly, the fur on his back raised. The dog continued to bark until David crossed to him and took him by the collar, stroking him. The animal was quivering beneath his hand, and little wonder, he thought. Of the other dog, Sabre, there was no sign. In fact, as David stood in the yard, noticing how cool the air had now become, he realized how quiet the dogs had been when the family had first come blundering into the house over fifteen minutes earlier. They normally barked at the slightest strange sight or sound, but both had been conspicuous by their silence until now. It was Julie who next emerged from the pub, Sabre at her heels.

'Has it gone?' Julie asked, looking around.

David nodded slowly, almost fearfully, as if any indication of the object's departure would immediately bring it back into view.

'What was that noise we heard earlier on?' she wanted to know.

'The buzzing?' he asked.

'No. There were some . . . bangs, explosions.'

'It must have been the fuses. That . . .' He hesitated to use the word. 'That . . . spacecraft must have overloaded the system.'

Spacecraft.

The word still stuck in his mind like a splinter. It still retained a certain unreality despite what he had witnessed that night. He scanned the dark sky once more then led the way back inside and bolted the front door.

'The phone's working again, David,' Caroline told him. 'I'll call the police.'

David hesitated for a moment, remembering what had happened at the Hughes' farm, but then he nodded and Caroline began dialling. David made his way towards the cellar, checking that the children were all right before he did so. Both were still shaken, understandably, but they had stopped crying and the sight of the dogs, strangely enough, seemed to have heartened them somewhat. He asked Julie too if she was feeling OK. She nodded, smiled weakly at him, and watched him as he walked through the bar towards the cellar door. He slipped the torch out of his pocket and opened the door, careful not to trip on the first darkened step.

Moving quickly and carefully, he reached the fuse box and opened it, aware of the pungent smell of burning in the air. It was mixed with the trace of ozone, too. David undid the fuse box and pulled it open.

Every fuse in it was burned out.

Not just burned but turned to charcoal, blown apart as if some massive force had been pumped through the wiring. He prodded one with his finger, swearing when he found that it was still hot. David reached into his pocket and rummaged around until he found his pen-knife. Then, carefully, he removed the incinerated fuses and replaced them with new ones which he took from a wooden box nearby. When every one was in position he pulled the switch to activate them.

'Try the lights,' he called to Caroline, who had appeared at the top of the stairs. She flicked the switch ahead of her and the cellar was bathed in a cold white glow. She did the same in the bar.

'I called the police,' she said as David made his way back up the stairs.

'Good,' he said. 'Perhaps this time they'll do something.'

By the time the police arrived, the family were in the sitting room drinking tea and eating sandwiches which the two women had prepared. As the first knock came on the front door King and Sabre began barking. David silenced them then went to answer the knocking.

He opened the door and almost stepped back in surprise. Constables Bryant and Vincent stood facing him.

'Hello again, Mr Ellis,' said Bryant.

David nodded and ushered them in.

'What can we do for you this time?' asked Vincent and David wondered if he could detect a slight air of sarcasm in the young constable's voice.

He led them through into the sitting room. 'It happened again,' said David, flatly.

'What did?' asked Bryant.

'The spaceship, the light, whatever you want to call it. It was here. It followed us and then it hovered outside.'

Bryant sighed softly. 'The same light you reported earlier?' he asked.

David nodded.

'But this time it started to dive at us,' said Caroline. 'It looked as if it wanted to kill us.'

The two policemen exchanged brief glances, then Bryant looked at Caroline.

'It wanted to kill you,' he said.

'Look, don't try to humour us, constable,' David snapped. 'That bloody thing, whatever the hell it was, looked as if it was going to collide with the car. The engine heated up, the whole *car* heated up. We were lucky to make it here.'

'And what happened when you *did* get back?'

'It hovered outside, then just disappeared, the same as before.'

Bryant nodded.

'Didn't anyone else report seeing anything strange tonight? Any lights or things in the sky?' asked Caroline.

'Not that I know of. I think we would have heard by now if they had.'

There was a long silence, then David got to his feet again. 'Earlier on,' he began. 'You said you couldn't do anything without physical proof of the sighting.'

Bryant nodded.

David crossed to the mantelpiece and picked up three of the burnt out fuses. He handed them to Bryant.

'I took those from the fuse box down in the cellar,' he said. 'The rest were like that.'

The family watched as the constable examined them and then passed them to his companion.

'What does that prove, Mr Ellis?' said Vincent, peering at the blackened lumps.

'The fuses blew while the craft was hovering outside the pub,' David told him. 'I think the added power caused some kind of overload.'

'That's not proof of the sighting, though,' said Vincent, defiantly. 'All this proves is that your electrical system overloaded. It could have been a short circuit, or anything. These,' he held out the fuses almost accusingly, 'don't prove anything.'

David rubbed his eyes and exhaled wearily. 'You don't want to believe us, do you?' he said.

Bryant shuffled uneasily in his chair. 'It's nothing to do with that, Mr Ellis, but like I told you before, we can't really do anything to help you.'

'So what *is* it going to take before you investigate these sightings?' David demanded.

'So far, only you and your family have seen anything,' said Vincent.

'And Alan Hughes,' David reminded him.

The older policeman nodded. 'That's six people.'

'How many do you need, for Christ's sake?'

'There was a spaceship out there,' said Caroline, pointing in the direction of the window. 'The same one we saw earlier. Only this time it got closer to us. It was trying to kill us.'

'Don't you think you're over-reacting slightly, Mrs Ellis?' Vincent said haughtily.

'You didn't see it,' David snarled.

There was a long silence, broken only by the ticking of the clock on the mantelpiece. David paced agitatedly back and forth, hands clasped behind his back. Vincent laid the burned out fuses on the coffee table nearby and wiped his palms together.

Bryant got to his feet. 'I don't know what else to say to you, Mr Ellis,' he said.

David didn't answer. 'And if this happens again?' he finally said.

'If only you could get a photo or something, any kind of proof, then perhaps . . .' He allowed the sentence to trail off.

David nodded but didn't look at the policemen. He turned his back on them and walked across to the bay window, gazing out into the night. Julie showed the uniformed men out and the family heard the sound of the car pulling away.

David shook his head. 'What the hell does he expect me to do? Walk around carrying a bloody polaroid just in case the damn thing appears again?'

Caroline crossed to him and touched his arm. It felt warm to the touch, the skin slightly reddened in places.

'Let's go to bed,' she said. 'It's late.'

'Thank God the kids haven't got to be up for school,' he said, managing a slight smile.

Julie led the two children up the stairs with a promise that she would sit with them until they fell asleep. Both were understandably shaken by the happenings, Michelle in particular. She had hardly spoken a word since they left the Hughes' farm.

'If it was a spacecraft of some kind,' said David, slumping into one of the leather-upholstered armchairs beside the fire, 'I wonder what it wanted?'

'You sound as if you doubt it, now,' said Caroline, sitting on the arm of the chair.

He reached out and slid his arm around her waist. 'I would have doubted it to begin with. I would have laughed the whole thing off if someone had told me it was going to happen to me.' He chewed his lip contemplatively. 'Now I don't know what to think.' He massaged the bridge of his nose between thumb and forefinger, aware of a slight tingling sensation around his eyes and cheeks. Probably caused, he reasoned, by the exposure to the bright light. A dull gnawing ache was beginning to grip the back of his neck. He slapped

Caroline's thigh gently.

'You're right, we should go to bed.'

He flicked off the lights in the sitting room, then walked through into the bar, checking that everything there was in order, resisting a quick look out of the window. Then he turned and headed back the way he had come up the stairs.

David lay on his side, staring at the glowing green hands of the alarm clock. They had just reached 3.26 a.m. Despite the physical and emotional exertions of the evening he could not seem to find respite in sleep. He felt hot, his body felt as if it were wrapped in steel wool. Every time he moved it hurt. He rolled onto his back and stared at the ceiling. Beside him, Caroline slept. She had moved closer to him, but David had moved away, the added heat from her body making him feel more uncomfortable. Now he reached up and brushed away the film of perspiration which had formed on his face and forehead.

For what seemed like hours he lay there like that, but finally, unable to stand it any longer, he swung himself out of bed and walked across to the window. His pyjamas were sticking to him and he pulled them off, tossing them in a sodden heap beside the bed. The tingling which had so far been confined to his face, now seemed to have spread over his whole body. He pushed open the window, allowing the cool air to wash over him. Sucking in deep lungfuls he closed his eyes and leant on the window-ledge. The air smelt of grass and damp earth and David inhaled the heady aroma, enjoying the feel of the breeze on his skin. He looked up, scanning the heavens.

Several thick banks of cloud had drifted into view, propelled by the breeze, and after a moment or two David returned to bed. He felt a little better, and in another fifteen minutes he was asleep.

Outside, high above the Derbyshire countryside, the glowing iridescent ovoid emerged from behind the banks of cloud and hovered. It remained motionless for more than ten minutes, then disappeared once more.

Five

July 23rd

'David.'

The voice seemed to be floating on air.

'David.'

He heard it but did not stir. The sound of the alarm clock died away as Caroline reached across him to turn it off. She shook him once more, forcefully this time, in an effort to rouse him.

'Are you all right?'

He rolled slowly onto his back and groaned softly.

'What time is it?' he asked, pressing both hands to his face.

'It's time we were up,' she told him. 'Don't forget we've got a pub to run.' Caroline swung herself out of bed and pulled on her dressing gown. David lay where he was, listening to the sounds of running water from the adjoining bathroom as she showered. He sat up, wincing as two sharp needles of pain bored into his temples. His eyes felt as if someone were trying to push them out of their sockets from the inside, and as he sat on the edge of the bed he rubbed a hand over his cheeks. They itched something wicked. The flesh felt rough and bumpy and David got up and crossed to the mirror on the wardrobe door. He swallowed hard as he studied his reflection.

His forehead, cheeks and neck were covered by dozens of small spots, like a rash, but the hives were slightly larger. They were on his forearms too. David touched the reddened area on his cheek tentatively, unable to resist scratching it.

The action only made the skin redder so, difficult as it was, he refrained.

Caroline returned from the bathroom, a towel wrapped around her. She watched David for a moment then caught sight of his reflection.

'What's the matter with your face, David?' she asked, drying herself.

He didn't answer.

'David.'

'What?' he snapped.

'Your face . . .'

He cut her short. 'I know.' There was a harshness to his tone which was quite unnatural. 'Have we got any calamine lotion?'

She began slipping on her clothes, telling him that there was some in the bathroom cabinet. As he passed her she looked more closely at his face. His eyes were narrowed in pain, but it was the rash which really caught her attention. It looked like an allergic reaction of some kind, but from what she couldn't imagine.

'Do you feel OK, David?' she asked him, watching as he filled the sink with water and splashed his face.

'I'm all right,' he said, tersely. 'I've got a bloody headache, though.' He touched the bumps on his skin cautiously. He took two pain-killers for his headache then reached for the calamine lotion and began applying it carefully.

'I'll see you downstairs,' Caroline told him. 'Do you want breakfast?'

'No.'

She paused for a moment, wondering whether or not she should say something. He was never as irritable as this in the mornings. As long as she had know him he had always been cheerful when he got up, whether he felt rough or not.

'Why don't you stop in bed for another couple of hours?' she suggested. 'You might be able to sleep that headache off. Julie and I can manage in the pub.'

He spun round. 'I told you, I'll be OK,' he snapped.

Caroline left the bedroom and made her way downstairs to

the kitchen where Julie and the children were already seated at the kitchen table eating bacon and eggs.

'Where's David?' Julie asked.

Caroline frowned and shook her head almost imperceptibly, a movement which she wanted neither of the children to see. Fortunately, they didn't.

'He's on his way,' she said, then went on to explain about the headache and rash.

Julie understood that there was something wrong and got up to pour her sister a cup of tea. It was all she wanted.

'Is it all right if we take Sabre out with us this morning, Mum?' asked Dean, stuffing bacon into his mouth.

He seemed as bright as usual and Caroline was relieved to see that the incidents of the previous night didn't seem to have bothered him. Michelle, on the other hand, was a little more subdued. Nevertheless, the two children seemed fine. Both had eaten hearty breakfasts, and Michelle had given two pieces of bacon to King, who was stretched out on the floor beside her. The other dog was in the bar, padding around amongst the tables and chairs.

It was another fifteen minutes before David joined the family, and as he walked in, Dean pointed to his father.

'What's wrong with your face, Dad?' he asked.

'It's nothing,' said David, sitting down and reaching for the daily paper which lay nearby. He accepted the tea and toast given him without acknowledgement, barely glancing at the others seated at the table. There was an awkward silence, broken by Dean.

'Dad, can we take Sabre out with us later on?' It was the beginning of the school holidays, and Dean had a lot of time to kill.

David apparently didn't hear.

'Dad . . .'

'What?' he snapped, lowering the paper and staring at his son.

The boy blanched visibly beneath his father's malevolent gaze and repeated his question more quietly.

'Yes, take him,' said David and returned to his paper.

Julie and Caroline exchanged puzzled glances.

Breakfast was completed in silence. Julie and Caroline finished the washing up, the two children and Sabre left the building for the hills and woods outside and David wandered into the bar to prepare for opening time.

He and Caroline had owned and run the pub for over fifteen years. It was a large, stone building with a low ceiling in the bar and a genuine oak counter which David usually polished before opening every day. This time he didn't bother. The pub stood about two miles outside Matlock itself, surrounded by rolling Derbyshire hills which were dotted every so often with outcrops of granite. The main road into the town lay three hundred yards down a narrow dirt track to the left of the building. Behind the pub there was a spacious garden where in summer metal chairs and tables stood, each one adorned with a brightly-coloured umbrella. Despite being somewhat off the beaten track, the pub did excellent business and many locals trekked out for both sessions of business. There were also numerous passers-by. But despite the briskness of its business, the pub never seemed crowded and it always carried that air of relaxed peacefulness about it. There was no juke-box. Darts or dominoes were the entertainments on offer and that seemed to suit everyone.

David checked the optics behind the bar, where necessary replacing empty bottles with full ones. He checked the beer barrels, tipping the froth into the slop tray until he was satisfied with the beer which flowed out. Usually he drank half a pint before opening up (his only alcohol all day), but not this time.

Caroline joined him in the bar while Julie got on with the business of preparing the home-made pies and pasties which had also helped to gain the pub its favourable reputation.

'Are you sure you feel OK?' Caroline asked him.

David nodded, pleased that the headache at least seemed to be receding slightly. At precisely 11.00 a.m. he opened up, and by 11.15, the first customers had arrived.

Six

Dean kicked the football, watching as Sabre ran after it. Michelle followed, riding her bike as best she could over the bumpy ground. The two children had plenty of friends in the town itself, but today they had decided to play on their own. They got on remarkably well for brother and sister and there was little of the animosity which usually surrounds the relationships of ten- and eleven-year-olds.

Despite the early hour it was already hot, the sun beginning its climb into the cloudless heavens. Only once did Michelle glance briefly at it and remember the glowing yellow 'egg' of the night before. In the green and warm surroundings, the horrors that they had endured seemed all the more distant. Nonetheless, she could not fully shake the visions from her mind, and as a plane passed overhead, leaving a thick vapour trail, she shuddered involuntarily.

The pub lay about half a mile behind them now, bathed in the bright sunshine. The hills grew steeper the further away they got and trees began to grow in greater abundance, eventually gathering into a thick copse which nestled in a dip between two of the undulations. Michelle got off her bike and pushed it to the top of the hill. From this high vantage point she could see the roof-tops of Matlock itself in the distance.

Dean was still kicking his ball about, chuckling as Sabre leapt about after it. The boy gave it a particularly hefty boot and it went flying down the hill, bouncing towards the large copse of trees. Both children chased after it, laughing as the dog skidded on the grass in his attempts to catch the ball.

The Alsatian stopped abruptly at the bottom of the hill.

The ball had rolled into the trees and effectively out of

sight, but instead of pursuing it, the dog merely stopped in its tracks and stared into the woods. As the two children caught up, Sabre backed off a yard or two and, still facing the trees, began growling.

'What's wrong, boy?' asked Dean looking first at the dog and then at the trees.

'He looks frightened,' Michelle added.

The dog barked loudly, ears pricked up, the hairs rising on the back of his neck.

Dean took a step towards him and the animal spun round, lips drawn back to reveal long canine teeth. Sabre growled at him and the boy moved away, allowing the dog to keep up his vigil which he now did, returning his attention to the trees.

'Perhaps there's a badger or something in there,' suggested Michelle.

'If there was he would have probably chased it,' said Dean, now more concerned with the maddened dog, which had begun barking more vigorously, his attention riveted to the woods. The boy waited a moment longer then advanced towards the copse. 'I suppose I'll have to find the ball.'

As he moved towards the nearest of the trees, the dog went berserk. Barks and growls were fused together into one cacophony of sound and the animal reared up on his hind legs, teeth bared in fury and fear. Fear? It was backing off still, despite this display of fury, and the further into the woods Dean moved, the more vehement the dog's actions became. He was almost slavering now, his barks loud and piercing. Finally, he stopped his backward retreat, turned and ran as fast as his powerful legs would carry him, back up the hill and out of sight.

Michelle watched the terrified dog disappear, then looked warily at Dean.

'We'd better find the ball,' she said. 'Then perhaps we should leave here, go and play somewhere else.' There was a note of anxiety in her voice.

'Why?' asked Dean, pushing his way further inside the enveloping gloom of the trees. The green canopy cut off a large proportion of the strong sunlight. Here and there shafts of the golden light speared through to illuminate the mossy floor of the copse. Michelle followed him, looking around nervously. It was she who was the first to notice.

'Listen,' she said, stopping momentarily.

Dean seemed too intent on finding his lost ball. He finally spotted the orange object about ten yards away and was about to retrieve it when Michelle caught hold of his arm.

'Listen,' she insisted.

'I can't hear anything,' said Dean, anxious to get to his football.

'That's what's wrong. There's no sound. Not even any birds.'

The two children stood quietly, straining their ears for any sound, wanting, needing to hear it.

The copse remained quiet.

Dean finally tired of standing around and scuttled off to fetch his ball. Michelle merely stood and gazed around her. Perhaps the birds in the woods were being quiet, she told herself.

'Michelle.'

The shout sounded thunderous in the silence and she spun round, heart hammering against her ribs. Dean was standing with his back to her, arms hanging limply at his sides, the football apparently forgotten at his feet. Michelle ran to him and, together, they stared at what lay before them. For long seconds they stared, not quite able to comprehend, but nevertheless feeling the icy hand of fear clutch at the backs of their necks. Dean moved backwards slowly and Michelle followed his example.

'We'd better tell Dad,' she said.

Her brother could only nod, his eyes still riveted to the sight before him. As he stepped backwards he tripped and fell. Michelle helped him up and both of them looked down, wondering what it was he'd fallen over.

When they saw what it was both of them ran like the wind, leaving the wood behind as quickly as possible.

'You look a bit under the weather, Dave,' said George

Fielder, looking at David over the rim of his pint pot.

David shook his head.

Fielder was one of the pub's regulars. He ran an undertaking business with his two sons in Matlock and he had been coming to the pub ever since it opened. Most of the regulars sat at the bar; casual visitors usually frequented the tables or, on days such as this one, sought the peacefulness of the garden. The pub was busy; David, Caroline and Julie serving behind the bar as well as dispensing food where needed.

The cuckoo clock above the bar had just signalled that it was noon.

'You're a bit old for acne aren't you, David?' said a younger man who was sipping at a whisky. He grinned and indicated the red blotches on the barman's face.

'It's a rash,' David snapped.

'Probably an allergy,' Caroline added, seeing the surprise on the younger man's face when David turned on him. Her husband's irritability didn't seem to have worn off even if his headache had.

David disappeared down into the cellar to fetch a new barrel of bitter and George Fielder motioned Caroline closer.

'Dave's a bit off today, isn't he?' said the undertaker. 'He's not usually like this.'

Caroline apologised for her husband and explained once more that he wasn't feeling too good. She wondered whether she should mention what happened the previous night. Even if it was only to find out if anyone else had seen the mysterious light. She surveyed the faces at the bar. Should she ask? She finally decided not to. It had been impossible to convince the police, so what would these people think? Surely if any of *them* had seen anything they would have said so by now. Or, she reasoned, were they too afraid of ridicule? Eventually she chose not to mention the lights or the craft.

David returned from the cellar, and as he did so, Dean and Michelle appeared from the sitting room. They had come in the back door, through the kitchen.

'Dad,' Michelle called.

He looked round. 'What do you want?'

'We want to talk to you for a minute.'

'I'm busy. Can't it wait?'

'Please, Dad.'

He muttered something under his breath and wandered off to join the two children who ushered him into the sitting room.

'I thought I told you never to bother me while it's busy,' David said, reproachfully.

'We found something in the woods,' said Michelle, pointing in the direction from which they'd come. She went on to explain what had happened. About the dog, about what they'd seen.

'We thought we'd better tell you, Dad,' Dean added.

David thought for a moment, looked at both of his children, saw the fear in their eyes. He asked again where they had been and they told him.

'You stay here until I get back,' he told them. 'Don't mention this to anyone.'

They both nodded vigorously.

He left by the back door, passing Sabre who was lying on his belly, whimpering. David gave the dog a cursory glance and set off towards the hills which masked the small copse. As he reached the top of the hill he looked back at the pub. It seemed so distant, he felt as if he were the only thing alive in the area. He made his way down the reverse slope, the woods drawing closer as he did so. When he reached the bottom he slowed his pace, approaching the trees cautiously. Michelle had been right, there was no bird song. No sound at all. Only the soft fall of his own footsteps on the grass.

It felt hot inside the wood, as if the canopy of leaves had trapped the sunlight, creating a greenhouse effect. Perspiration popped out on his forehead and trickled slowly down his face.

Just ahead of him he caught sight of Dean's football. He knew he was close to the place the children had mentioned.

David sniffed the air.

There was an acrid, biting odour to it. Rather like . . . He tried to place the smell. Like . . .

Burnt wood.

'Jesus Christ,' he murmured, gazing at the sight before him.

Trees and bushes all around had been seared by some incredible source of heat. The firs and pines were little more than charcoal, even the earth itself had been blackened. But most remarkable of all was the massive circular patch of burnt ground which lay before him. David guessed that it must be somewhere close to twenty or thirty feet in diameter, an almost perfect circle. As he walked onto the patch he heard the ashes crunching beneath his feet, but rather than the fear which the children had felt, he experienced a kind of serenity. It seemed as if he were walking on air, things around him became blurred and indistinct. His movements became slower until he found it almost impossible to lift his legs, it was as if all the energy were being drained from him. His legs, which had felt like cotton wool, suddenly felt like lead weights. David winced at the effort of walking, his feet crunching heavily on the scorched earth beneath. He headed back to the perimeter of the circle but it seemed to take him an eternity to reach it. The blackened trees lowered over him like dark sentinels, watching his clumsy steps almost with glee. He finally made it to the edge and fell forward. It was like suddenly escaping a strong magnetic field. Free of its pull he lay still on the damp earth, sucking in lungfuls of moist air, but there the ever-present smell of burned trees and plants enveloped him.

David finally managed to haul himself upright, his head spinning. He felt sick but steadied himself for a moment and the feeling passed. He leant against a tree to recover his wits.

Whatever had made that scorch mark was large and he couldn't begin to imagine the power and heat which must have been involved, but somehow, his brain was clouded. He tried to think, to reason, but found it impossible. He staggered away from the tree and stumbled, almost falling as his foot sank into a deep indentation before him.

It was a few seconds before he realized that it was a footprint.

Twelve, fifteen inches long. He could only guess. Whatever had made it was certainly not a man. The print was far too large and, judging by the depth which it sunk into the earth, the creator of that footprint had been heavy. Very heavy indeed.

There were others.

They led away from the burnt patch for about ten yards then stopped and, from what David could see, their creator had not left a corresponding set to show his return. It was as if he had merely back-tracked, walking in his own footprints.

David followed the prints back to the scorched circle again, once more experiencing that peculiarly disquieting 'pull'. He lowered his gaze, closing his eyes momentarily.

The silence was deafening, only the roaring of his own blood in his ears was audible.

For long seconds he stood there, then, by what seemed like a monumental effort of will, he turned his back on the blackened patch and set off out of the woods. He found that he was shaking slightly. What the hell had happened there? Had the craft actually landed? And the footprints. If they were twice the length of his own then how big was their owner? The questions tumbled around in his mind, none with an answer.

He eventually reached the hillside again and dropped to his knees, glad to be breathing fresh air once again. He rolled onto his back and gazed up at the sky. It was clear and blue, cloudless. He remained like that for about five minutes, then got to his feet and plodded back towards the pub, already determined not to mention anything about the incident to Caroline just yet.

As he made his way back the vision of those huge footprints stuck in his mind.

They had stopped after ten yards.

Next time, and he felt sure there would be a next time, God alone knew where they would stop. At the top of the hill? Outside the pub itself?

Despite the warmth he shuddered.

Seven

The hands on the cuckoo clock had crawled round to 11.25 p.m. by the time the last customer left the pub. Caroline breathed a sigh of relief and locked the door.

Behind the bar, Julie and David were busy washing and drying glasses, replacing them in the correct position in readiness for the next day's business.

Caroline had put the kids to bed about two hours earlier and she decided to check on them when everything was cleared up for the night. Now she lifted chairs onto the tables and emptied the ashtrays into a large bin which stood in the middle of the floor. David didn't speak as he continued with his task. In fact, Caroline realized, he had hardly said above twenty words since dinner time. His earlier irritable mood had darkened into a brooding silence which made her feel even more uneasy than his initial hostility had.

For his own part, David was still wondering whether or not he should mention what he'd seen in the copse over the hill. Not just the footprints but the peculiar feelings he'd experienced. How was he to explain it?

'I think that spaceship or whatever the hell it was landed last night,' he said, the words flowing out effortlessly.

Both Caroline and Julie stopped what they were doing and looked at him. He continued washing the glasses.

'How do you know?' asked Caroline.

'In the woods,' he told them. 'I saw some trees. They'd been burnt to cinders and there was a patch of earth too. Everything was burned.' He explained about the footprints as well.

'Oh my God,' said Julie. 'What can we do?'

'How the hell do I know?' David snapped.

'Do the kids know about this?' Caroline wanted to know.

'They were the ones who found it,' David told her.

There was an uneasy silence broken by Caroline. 'How did Dean and Michelle take it?' she wanted to know.

David shrugged. 'They seemed frightened, but . . .' He let the sentence trail off. 'It was Sabre who really reacted to the landing.'

'You sound so sure that it was a spaceship,' said Julie.

He rounded on her. 'Well what else could it have been?'

Another silence descended, this time broken by the low growling of one of the dogs. It was King, the larger of the two Alsatians. The animal had been padding around the bar, sniffing at crumbs on the floor but now he stood facing the front door, standing stiffly there, ears pointed forward. His growls grew louder. All three of them looked at the dog which had now brought his shoulders forward as if he were bracing himself.

There was a flash of bright light from outside the window and King began barking.

'What was that?' Julie pondered aloud, as another white flash illuminated the bar.

'It could be lightning,' said Caroline. 'Maybe we're in for a storm.'

There were two more bright explosions of light, and this time the three of them crossed to the windows and looked out.

King, by this time, was barking madly and his wild yelps had been joined by Sabre's, from the sitting room. Both animals were making a fearful racket. David tried to silence King, but as he turned the animal bared his teeth. David stood still for a moment, watching the dog which was now barking at *him*.

The flashes of light continued.

'Where's the thunder?' Julie murmured to herself, peering into the darkness outside. She recoiled slightly from the next brilliant streak which burst across the sky. The flashes were soundless, unlike lightning, and Caroline noticed that they now seemed to be coming more regularly, almost at the same time. Measured seconds apart. The dogs were still growling and barking and David dashed through into the sitting room to find Sabre up on his hind legs looking out of the bay window.

If anything, the light seemed more powerful at the back of the pub and each fresh explosion illuminated the garden.

He paused for a moment, watching the dog, then he hurried through into the bar once more. Caroline nearly bumped into him as he passed.

'Where are you going?' he asked.

'To see if the kids are all right,' she told him, disappearing upstairs.

The flashes of light continued at short intervals and David and Julie could only stand helplessly as the dogs continued to bark wildly at the blinding illuminations.

The lights over the bar began to flicker and die.

Julie shot David an anxious glance. He dashed behind the bar and checked on the bulbs which were now blinking on and off, becoming dimmer each time.

The entire building was suddenly plunged into darkness.

The dogs' growls turned to placid whimpers, and in the silence David and Julie heard the sound of footsteps on the stairs as Caroline returned.

'The back lawn,' she said, quietly, her face pale.

They followed her through into the sitting room and looked out of the window.

The craft, now all too familiar a sight to them, was actually grounded on the rear lawn of the pub. It radiated its usual yellow glow, but tinged with a more powerful purple light this time. David guessed that it must be no more than two hundred yards from the building.

Without thinking he reached for the nearby phone.

It was dead.

He threw the receiver down angrily.

'We've got to get help,' he said.

'But how, if the phone's out?' said Caroline, her voice a mixture of desperation and pure terror.

'I'll have to drive into Matlock,' he said, flatly.

Caroline gripped his arm.

'David, you can't go outside, not while that thing is there.'

He pulled free of her. 'We don't have any choice,' he rasped, fear now creeping into his own voice. 'Lock all the doors and windows and keep down. Stay hidden. Come and lock the front door behind me.'

Caroline hesitated but he strode through into the bar.

'Now,' he insisted, his hand resting on the door. He slid back the bolts and opened the door a fraction. 'Come on,' he shouted.

She ran across to join him. 'Please, David . . .' The sentence trailed off, her voice cracking.

'Lock it behind me,' he said and dashed out into the gloom.

Caroline did as she was told then stumbled back into the sitting room where Julie was crouched behind one of the armchairs. The entire room was filled with a piercing white glow, and both women could only cower like terrified children, wondering what was going to happen next.

David was thankful that he had not parked the Volvo in the garage at the side of the pub; that would have meant going close to the garden to reach it. The car was standing a few yards from the front door and he sprinted towards it, not daring to look behind him. He tore open the door and slid behind the wheel. The engine started first time and he stuck the vehicle in gear, pulling away quickly. He glanced into his rear-view mirror and saw that the pub was in complete darkness, only the vile yellowish glow of the landed craft lit the night.

He floored the accelerator and headed for Matlock. As he drove he squinted into the night. Was he being followed? He gripped the wheel tighter. His mind was in a turmoil, he knew that the only way to get help was by driving into town but he dared not think about what was going on back at the pub while he was away. Should he have left his family alone? The question tortured him.

A sign flashed past which announced, 'Matlock one mile'.

He tried to coax more speed from the Volvo, but to his horror he found that the car was skidding, the engine bucking more often now.

David looked down and saw that the needle of the petrol gauge was deep into red. Another few hundred yards and he would be out of fuel. The thought made him panic. He was on a main road, granted, but it was a deserted main road. What if the craft had left the pub and was scanning the ground for *him*? He was alone and helpless. He banged the steering wheel angrily but the car merely stalled, and with a final grating sound from the engine, it rolled off the main road. David leapt out, almost falling on the tarmac. He looked behind him, praying that he wouldn't see the yellow 'egg' in the sky, but somehow wishing that he could. At least that would mean that his family were safe.

He began running, fired with speed he never knew he had. A speed born of fear.

His breath was coming in gasps now, and every few yards he looked over his shoulder, scanning the night sky for signs of light or movement. He hadn't far to go now, the police station was close, but it seemed like a marathon, and for each step he took, it seemed further away.

There were lights behind him.

'No, please God, no,' he whimpered, hardly daring to look round. He kept on running, his feet pounding on the tarmac.

The lights were getting brighter, pinpointing him on the lonely road.

He felt like screaming.

The lights of the first few houses were visible ahead of him while those behind drew ever closer.

David finally found the courage to look over his shoulder.

The car which had been behind him swept past, the driver casting a cursory glance at the unkempt man who was running on the roadside. David felt his knees buckle and he thought for a moment that he was going to fall. The fact that his pursuer had been another car and not the glowing spacecraft did little to help his anxiety. Only when he finally reached the police station did he feel secure.

He burst through the front doors and up to the desk where a portly man in a sergeant's uniform was drinking tea from a large mug. The policeman immediately put down his tea and looked at David.

'Help me,' he gasped. 'My family are in danger.'

Sergeant Phillip Gregory wiped his moustache with his thumb and forefinger and studied David closely.

'You're from The Horseshoes aren't you? That pub . . .'

David cut him short. 'My family are in danger, for God's sake. I need help now,' he gasped.

Constable Bryant emerged from a back room and saw David.

'*You* know what's been happening,' David said, pointing at the constable. 'You've got to help me.'

'Just calm down,' said Gregory. 'Tell me what's happening.'

'The craft, it's landed,' David blurted out.

Gregory looked puzzled. 'Craft? What bloody craft?'

'I'll take care of it, sarge,' said Bryant. 'Come on, Mr Ellis, I'll drive you back home. You don't look in a fit state to do it on your own.'

'My car ran out of petrol about half a mile up the road,' David gasped.

'Take one of the Pandas,' said Gregory, still eyeing David suspiciously. 'What the hell is he on about anyway?'

'I'll explain later,' Bryant said, guiding David out of the building towards one of the waiting police cars. They both climbed in and the constable started the engine.

'Hurry, for God's sake,' said David.

Bryant pulled away, flicking on his lights as he did so.

'You wanted physical evidence of what we'd seen,' David rasped. 'Well, now you'll see it.'

Bryant looked across at his passenger. He certainly did seem distraught, but three sightings in two days . . . The constable was beginning to wonder what the hell Ellis was playing at. Physical proof. He smiled softly to himself.

★   ★   ★

Caroline and Julie were still crouched on the floor, Caroline in particular was trying to quieten one of the dogs which was continually whimpering. The other was in the bar somewhere, but apart from the sounds the dog was making, the pub was in silence. The light from the landed craft burned as brightly and from where she crouched, Julie could see the top of it from the bay window.

'What is it doing?' Caroline asked, not really expecting an answer.

There was a scream from upstairs.

'Oh my God;' Caroline moaned. She was on her feet in seconds and dashing for the stairs, stumbling in the darkness. She blundered into Michelle's bedroom and saw that the child was standing at her window looking out onto the lawn. She did not move, her attention riveted to something other than the space-craft. Caroline hurried across and took her in her arms, squeezing her tightly.

'Look,' Michelle whispered, pointing.

It was all Caroline could do to stifle a scream.

Standing on the lawn, bathed in the unearthly glow from the ovoid, was . . .

What was it? Her mind reeled as she tried to comprehend just what the hell it was.

The shape was humanoid, but tall, perhaps seven feet or more. It was difficult to make out any details of the figure because of the blinding radiance coming from the craft behind it. But as the figure moved forward, Caroline could see that it was glinting. No, she corrected herself, it was glistening. It didn't appear to be wearing any clothes, its body sheathed instead in a thick film of slime. It moved slowly away from its craft and, with horror, Caroline realized that it was heading towards the house.

She took Michelle by the hand and led her out onto the darkened landing where they were joined by Dean. Michelle was crying softly and she gripped her mother's hand tightly. Carefully they descended the stairs.

There was a loud bang on the front door.

Caroline froze as it was followed by another. Then

another.

She saw Julie emerge from the sitting room and head towards the door.

'Don't open it,' she screamed at her sister as another impact rattled the door in its frame.

Julie froze.

The banging stopped and, in the silence, only the sound of the dogs growling could be heard.

'What is it?' Julie asked, frantically.

Caroline had no answer for her. She hurried the children downstairs and, joined by Julie, they hid behind the bar. All of them, for the first time, became aware of a loud buzzing sound which seemed to fill the entire pub. Above all that, the banging came again.

'The back door,' said Caroline, her voice low and full of dread.

She looked at Julie. 'Oh God,' she blurted. 'I forgot to lock it.'

She was on her feet in seconds, moving through the sitting room, ducking low as she passed the window, and into the kitchen. She could see a dark shape silhouetted against one of the kitchen windows, the light from the craft behind it illuminating it. The infernal buzzing still sounded loud in her ears but now she had other things to worry about. The shape was moving towards the back door.

Caroline dashed across to the door and slid the bolts.

A second later there was a loud bang as something heavy impacted on the wood.

It was all she could do to stop herself screaming.

For interminable seconds she stood rooted to the spot, her eyes fixed on the back door.

The banging came again, harder this time it seemed.

Caroline tried to slow her breathing but she couldn't. Already she was wondering what to do should the creature break in. She and the family would have to run for it, there was no other choice. She tried to drive the thought to the back of her mind but it refused to be shifted and she clenched her fists tightly together as there was another powerful bang on the door. It was almost as if the being were testing the resistance before it finally smashed its way through the door, as if it were playing a little game with them.

And if it got in . . .

She ran back into the bar and got down on the floor again. The dogs, by now, were growling more fiercely, their attention directed towards the sound of the banging. King approached the kitchen but paused on the threshold, teeth bared.

Where was David? Caroline closed her eyes, pulling her children closer to her. Where was he? Oh, please God help us, she said to herself.

The banging ceased abruptly, only the persistent buzzing remained. For what seemed like an eternity nothing happened until the silence became even more unbearable than the pounding. What, Caroline wondered, was the alien plotting now? She hugged the children tighter to her and cast an anxious glance at Julie who was crouching, wide-eyed, beside her.

'What was making that noise?' she asked.

'Something came out of that spaceship,' Caroline explained.

'Did you see it?'

'Not properly.'

'It was a monster,' Michelle whimpered.

Still the silence. Only the ever-present buzzing. The light glowed brightly.

'What's it doing?' Julie gasped through clenched teeth.

There was a sudden powerful droning sound and the light began to fade. Caroline got to her feet again and moved cautiously towards the sitting room door. From her vantage point she could see the craft rising. With infinite slowness it left the ground, hovering about twenty feet up for about two minutes, then quicker than she could draw breath, it was gone. Vanished as if into thin air. She moved into the sitting room and across to the bay window. Of the craft there was no sign. Even high in the sky she could see no trace of the yellow object. Caroline let out a shaking breath and leant against the window-ledge for support. A moment later she was joined by Julie and the two children.

'It's gone,' said Dean.

Caroline seemed to forget what had just gone before and she embraced both her children. 'Are you all right?' she asked them.

Dean nodded. Michelle did not move. Caroline gripped her shoulders, gazing into her daughter's eyes. They were puffy and swollen from crying and there seemed little recognition in them as she looked at her mother.

'Michelle,' Caroline repeated. 'Are you all right?'

The girl nodded almost imperceptibly then, unexpectedly, collapsed in a heap.

Caroline screamed.

'She's only fainted,' Julie said, lifting her niece onto the sofa and, indeed, as she did so, the girl opened her eyes again. Dean watched the tableau with bewilderment.

'We've got to be sure that it's gone,' said Julie, peering into the gloom. She was making her way to the back door.

'No,' shouted Caroline. 'Wait until David comes back.'

David. Where was David, she wondered? What if something had happened to him? She felt like crying herself but knew that she must not. At least not in front of the children. She bit her lip and pulled some strands of hair from Michelle's forehead.

How long had David been gone? Fifteen minutes? Half an hour? An hour? Time seemed to have lost all meaning. She looked at the clock on the mantelpiece.

11.30.

Caroline frowned. It was later than that. The clock must have stopped. She checked her own watch.

11.30.

She got to her feet and walked into the bar. The cuckoo clock too was stuck at 11.30.

'Julie,' she called. 'What time is it?'

She heard a muttering from the other room.

'I don't know. My watch has stopped.'

Caroline continued to stare at the frozen hands of the clock.

Eight

It had begun to rain lightly by the time constable Bryant swung the Panda car into the driveway which led to the pub. Still shrouded in darkness, the building itself seemed to be just another part of the night, an extension of the hills and trees which rose on all sides of it. David felt his heart quicken when he saw that the lights were out. He fidgeted anxiously in his seat.

'I hope we're not too late,' he said.

Bryant didn't answer. He brought the car to a halt outside the pub and climbed out, walking slowly towards the main door. David beat him to it and banged hard.

'It's me,' he shouted. 'Open the door.'

'David?' came a voice from inside.

'Open up, I've got the police with me.'

He heard the sound of locks being unfastened and then the door swung open to reveal Caroline standing before him.

David pushed past her, scanning the bar for any signs of damage.

'Is everyone all right?' he asked, looking her up and down.

Caroline nodded. 'Julie and the kids are in the sitting room,' she told him. 'What happened to you?'

He told her about the car running out of petrol.

Bryant sauntered in behind him, closing the door after him. He was carrying a torch which he'd taken from the car.

'The fuses blew again,' said David.

The policeman nodded and followed him into the sitting room.

Once seated, Bryant got out his notebook and listened dutifully as the family relayed the details of the latest sighting. The spacecraft landing on the back lawn, the flashing lights, the clocks stopping and then even David listened in awe as Caroline described the emergence of the alien.

When they had finished the constable nodded wearily and slid the notebook back into his pocket.

'Aren't you going to telephone a report through?' David demanded. 'Get some more men out here?'

'Perhaps it might be an idea if you showed me where this . . . spaceship is supposed to have landed,' said Bryant.

'There's no suppose about it,' David snarled. 'The bloody thing was out there.' He crossed to the kitchen and wrenched open the back door, stepping out into the night. Bryant followed, and so too did Caroline, anxious to see if the creature had left any traces on the door in the course of its pounding.

It hadn't.

David led the policeman down the lawn to the spot where he had seen the craft and Bryant shone his torch over the area indicated.

There was a low rumble of thunder in the background.

'I can't see anything, Mr Ellis,' said the PC.

'I'm telling you, it was here.' David pointed at the ground beneath his feet.

Bryant shone his torch over a wider area but still could see nothing out of the ordinary. There wasn't a blade of grass out of place. The rain was falling more rapidly now and a fork of lightning tore across the sky. David jumped visibly as he saw the bright light which was followed, a moment later, by a thunderclap.

'Shall we get inside out of the rain, sir?' Bryant said.

'You don't believe me, do you?' David said. 'None of it.'

Bryant was about to speak when David's eyes lit up. He began walking across the garden towards the fence which separated it from the closest of the grassy fields nearby.

'Come on,' he said. 'You want proof, I'll show you.'

David ignored the rain as he led the policeman over the hills towards the woods beyond. Both of them nearly lost their footing on the sharp incline of the reverse slope and the heavens continued to dump their load on the earth. Every now and then a noiseless flash of lightning would whiplash across the sky followed by a low rumble of thunder and, each time, David would look up. As if he expected to see the yellow glowing ovoid reappear at any moment. The darkness, however, remained total.

They reached the bottom of the hill and he led the way into the wood where both men were, to a certain extent, shielded from the rain by the trees. Bryant flicked on his torch and shone it ahead of them, the powerful beam picking out a path through the undergrowth. He was beginning to wonder what the hell David was up to. Three times now he'd had to deal with reported UFO sightings in two days and this time it was aliens too. Bryant shook his head. He hadn't thought that David Ellis was a liar, nor any of his family for that matter, but what the policeman couldn't figure out was if there was so much of this sort of thing going on, why had no one else seen it and reported it? Admittedly, old Alan Hughes had backed up David's description but Bryant had known the farmer for years and he wasn't averse to a drop of the hard stuff every now and then. Bryant told himself that he should keep an open mind, but he was finding it increasingly hard to do so.

David stopped abruptly, pointing ahead of him.

'There,' he announced, directing the constable's attention to the large patch of scorched earth which he'd discovered earlier that day.

Both men walked onto it, the incinerated ground crunching beneath their feet. David stood still in the centre of the circle, his head spinning. Once more he felt that peculiar sensation, as if his entire body were made of lead. Even when he tried to talk it required a massive effort just to move his jaw.

'The spacecraft made this mark,' he said, his speech slow and almost slurred.

'How can you be sure?' asked Bryant, shining his torch over the burnt trees and bushes nearby.

'It landed here last night,' David told him.

'Did you see it land?' Bryant asked.

'Jesus Christ,' rasped David. 'What else could have made a mark like that? These trees have been incinerated. You tell me what else could have done that?' His tone was rising.

There was a rumble of thunder, almost like a challenge.

'Well?' he demanded, moving slowly away from the burnt area, dragging his legs with effort.

'Lightning,' said Bryant, flatly. 'I've seen trees struck before.'

'No,' David shouted.

'You must try and understand how this looks from my position, sir, I . . .'

David turned on him vehemently. 'It wasn't lightning that did this. Never. I'm telling you, a spacecraft landed here, the same one that my family and I saw this evening.'

Bryant didn't speak. He didn't really know what to say.

'I found footprints too,' David continued, searching the ground for the tell-tale tracks which he'd discovered. 'Shine your torch over here,' he told the policeman, pointing towards some bushes.

Bryant did as he was asked but, by this time, his patience was beginning to wear a little thin. Nevertheless, he followed David to the appointed place and watched as the pub owner dropped to all fours on the damp earth and began searching for something which he seemed convinced was there.

'Mr Ellis . . .' he began.

'There were footprints here,' David told him.

'Perhaps the rain washed them away,' said Bryant, wearily.

'They were *here*,' David told him, defiantly.

'Well they're not here now sir,' the policeman said, flicking off his torch.

There was a long silence, finally broken by David.

'I'm not doing this for a joke, constable,' he said, quietly.

'No, sir.'

'We saw a UFO. It landed here last night and at the back of our place tonight. Something, don't ask me what, got out of it. There was something here.'

There was another low rumble of thunder, followed by a bright flash of lightning and, in the cold white light, David's face looked ghostly to Bryant. He helped the pub owner to his feet and the two of them began walking back towards the pub. David looked behind him as they reached the top of the hill, glancing at the woods, then as had become his habit, at the sky.

Constable Graham Bryant climbed into the waiting Panda car and started the engine. He flicked on his windscreen wipers to clear the water from the glass then turned the vehicle and drove off. The pub became a distant image in his rear-view mirror. Its lights were on, David having fixed the blown fuses again. Bryant guided the car onto the road and headed back towards Matlock. There was little or no traffic around and he passed just one vehicle on his travels. The rain writhed like mist in the headlamps' powerful beams and Bryant decided to keep his wipers on.

He drummed the wheel with his fingers as he drove. He felt uneasy. Something nagged at the back of his mind. He was well on the way to convincing himself that what David Ellis and his family had seen had been some kind of illusion. The glowing lights especially could have been ball lightning and it *had* been stormy lately. Alien footprints. Creatures emerging from spaceships. He shook his head and frowned. Yet still, as he had reasoned before, there was nothing which the family could hope to gain from fabricating such an elaborate story. His logical training and experience told him that the whole thing was in their imagination, but a nagging doubt at the back of his mind persuaded him otherwise.

Flying saucers. UFOs. Aliens.

He smiled thinly to himself.

There was a sharp hiss of static as the two-way radio in the car burst into life. In the silence, it sounded thunderous. Bryant jumped in his seat, shocked by the harsh sound, then exhaling deeply, he reached for the hand-set. This bloody business about things from outer space was beginning to get to him. He smiled, wondering how the hell he was going to explain the events to Sergeant Gregory if he ever got around to asking. Gregory was short on imagination but totally devoid of a sense of humour.

'Baker One,' said Bryant.

'Bryant, where have you been?'

Talk of the devil, the constable thought, smiling to himself. The voice at the other end belonged to Gregory.

'Out at The Horseshoes, Sarge,' he said.

'I know that,' Gregory said. 'What's been going on?'

'Not much really.' Bryant thought about mentioning what he'd been told but decided to cut things to a minimum. 'Mr Ellis and his family, they thought they saw a spaceship.'

There was a moment's silence.

'A what?' Gregory demanded.

Bryant smiled. 'A spaceship.'

'I reckon he must have been drinking his booze, not selling it,' the sergeant said.

'Well, I'm on my way back now,' Bryant told him.

'Don't hurry. Just drive around the town for an hour or so, check everything's all right.'

'Will do. Over and out.'

He replaced the handset, puzzled at how hot the plastic had become. There might be a faulty connection in there somewhere, he thought; when he stopped he'd check it over. Then he realized how warm the entire car had become. He wound down the window, ignoring the rain which drifted in. It was stifling in the car.

A particularly powerful set of lights suddenly flashed in his rear view mirror and Bryant cursed to himself. Dip them, he thought, wincing. But the lights seemed to get brighter. They must be fog lamps, he thought, because they had a yellow tinge to them. The driver must have pretty bad eyesight if he needed an extra set of lamps. The rain was clearing now. Bryant slowed down a little and the light in his mirror grew brighter, glowing white now, but for the first time he noticed that there was just one ball of luminosity, not two separate lights. It seemed to be one enormous explosion of radiance.

Bryant slowed a little more and prepared to look over his shoulder.

There was a deafening burst of static from the radio followed by a loud buzzing sound.

The inside of the car was like a greenhouse by now, despite the open window. The constable looked down at the dashboard to see that his engine was getting hot, the needle on the temperature gauge rising steadily.

When he looked again, the light in his rear view mirror had gone. The road was in darkness again but for his own headlamps. No car had passed him and there were no turn-offs. He stopped the police car and looked behind him.

The road was empty.

For a full thirty seconds he remained like that, squinting out of the back window into the gloom, but he could see nothing. He turned round again.

Ahead of him, a little more than twenty feet above the road was a glowing yellow, egg-shaped object. Bright light radiated from it like heat from an electric ring. The object was motionless, as if held there by some giant invisible hand.

Bryant swallowed hard, his hand gripping the steering wheel tightly. Despite the warmth in the car he felt a cold chill race up and down his spine as he watched the glowing ovoid which was now beginning to rotate very slowly. Still in that one spot in the air, unmoving but for the slow rotation, it seemed to present itself for view and Bryant was able to see it clearly.

It was fully thirty feet across, curving smoothly at each end to form the egg-shape which David Ellis had spoken of. The policeman suddenly felt very stupid and very angry with himself for not believing what he'd been told and yet, even now, as he saw it with his own eyes, he found it hard to believe.

The craft moved off, away from the car and Bryant pressed his foot down on the accelerator, following it.

Eyes glued to the rapidly-moving shape he wondered whether or not he should report in to the station, tell them that the Ellises had been telling the truth, that there was a spacecraft. His mind filled with thoughts. Was it a spacecraft? Was there any possible way it could be a light aircraft of some kind? He shook his head. No one could fly a plane as low as twenty feet, especially not in terrain like this. Whatever it was it was being guided with infinite care and skill and it was moving without making any noise. All he could hear was that peculiar buzzing in his ears.

There was another loud burst of static from the radio, and this time, when he picked up the handset, no sound issued forth. He flicked the transmitter button but the set was dead. Bryant dropped it, his gaze moving once again to the object which was now increasing its speed. Determined not to lose it, he pressed down harder on the accelerator, the needle on the speedometer bouncing up to sixty. On wet, winding roads it was difficult driving at speed but Bryant didn't slow up, knowing that if the craft followed its present route, it would soon come to a stretch of straight tarmac. But, he reasoned, the glowing shape was not restricted to the confines of man-made roads and, indeed, a moment later, it veered to the left and over a field. There was no doubt about it, the thing was speeding up.

The policeman looked down briefly to see that he was now travelling at over seventy miles an hour. Yet still he tried to coax more speed from the Panda car.

The glowing shape flew noiselessly onward as if inviting pursuit.

The needle on Bryant's speedo touched eighty and he felt a shudder run through the car. His headlamps picked out a slight rise ahead but, so determined was he not to lose sight of the yellow ovoid, he didn't slow down.

The car hit the top of the rise doing eight-five. It hurtled into the air as if fired from a cannon and seemed to fly for interminable seconds until, with a bone-jarring crunch, it landed on another road. The wheels screamed as Bryant tried to regain control but it was useless. The Panda spun round twice before crashing into a wooden fence on the other side of the road. The craft stopped abruptly, hovering in the air above the Panda then, as Bryant watched, it shot off at a speed far in excess of anything within human comprehension.

The policeman could only sit stunned as the craft disappeared. It didn't leave a trace. No vapour trail. Nothing. He swallowed hard, trying to get his breath. His entire body was shaking and he closed his eyes for a moment, the vivid yellow afterburn of the object still bright on his retina.

The radio crackled once and he reached for it but there was no one on the other end of the line. He regarded the handset for what seemed like an eternity. Should he radio in what he'd just seen? He looked at the handset again, imagining Sergeant Gregory's reaction. Besides, as he himself had told the Ellises, what could be done without physical proof? And he, like they, had none. Bryant exhaled deeply and replaced the two-way, head bowed over the wheel.

He wasn't sure how long he sat there, he might even have blacked out for a moment, but eventually he started the car and reversed back onto the road. Fortunately, the Panda didn't seem to have suffered any damage so he wouldn't be forced into any explanations there.

No, he had decided that silence was the best and only policy. He would forget that anything had ever happened, that he'd ever seen anything.

But as he wiped the perspiration from his forehead, Bryant knew that was easier said than done.

He scanned the heavens one last time and shuddered.

Nine

July 24th

The morning was overcast, the grass still slippery from the previous night's rain. Clouds scudded across the sky. Warnings that there was to be no sunshine that day. Dean prodded at his breakfast, resigned to the fact that he and Michelle would have to stay inside until the weather got better.

David was looking at the paper, his half-eaten breakfast pushed to one side. Caroline and Julie were washing up prior to making the usual batch of pies and pasties.

'Where's Michelle?' asked Caroline, the question directed at no one in particular.

'Still in bed,' Dean said.

'I'm not surprised,' said Julie. 'After what happened last night.'

'Can we go out later, Mum?' Dean asked.

'If the weather gets better,' said Caroline.

'Stay away from those woods,' David said from behind his paper.

Dean didn't answer. He seemed surprised at his father's unexpected intervention.

David lowered the paper. 'Did you hear me?' he said. 'If you go out, keep away from the woods.'

Dean nodded, a perplexed expression on his face.

David dropped the paper on to the table and got to his feet, walking through into the bar where he began his usual chores which had to be completed before opening time.

'He's still the same,' said Caroline, softly. 'Ever since that first night . . .' She allowed the sentence to trail off. She looked round as if afraid that David might re-enter the room and hear her.

'Perhaps he should see a doctor,' Julie offered.

Caroline nodded. 'You know David, he won't go near a doctor unless he's dying.' She paused for a moment. 'It's just not like him to be so moody and irritable.'

'He'll come out of it, Caroline,' said the other woman.

They exchanged thin smiles and continued with their work.

'Are you going to wake Michelle?' Julie wanted to know.

Caroline shook her head. 'It's probably best if she sleeps.'

Dean pushed away his empty plate and climbed down from his chair. He was on his way to the sitting room.

'If you're going up to your room to play,' Caroline called after him, 'don't make too much noise, you'll wake your sister up.'

He muttered an affirmation and she heard his footsteps on the stairs.

'*You* look as if you could do with a few extra hours' sleep,' said Julie, studying her older sister's drawn appearance. There were dark rings beneath her eyes and her hair looked lank and lifeless. Quite the opposite to her normally radiant look. Julie herself felt tired but she had disguised the paleness of her appearance with make-up.

'I didn't sleep much last night, I must admit,' said Caroline. 'I don't think any of us did.'

'I wonder how all this is going to end?' said Julie.

Caroline looked at her. 'Don't make it sound so final for God's sake,' she said.

The two women looked at one another for a moment, then continued with their tasks.

There was a loud crash from upstairs and both of them cast anxious glances upwards.

Caroline turned and made her way towards the stairs. She found herself, unaccountably, running. There was another loud bang as she reached the landing. It came from Dean's room. She pushed open the door to find two large boxes upturned on the floor. Dozens of tiny plastic soldiers were spread out across the floor and Dean stood amongst them, an apologetic look on his face.

'What the heck are you doing?' said Caroline, trying to keep her voice low.

'Sorry, Mum,' he said, motioning towards the wardrobe from which the boxes had fallen. 'I was trying to reach something and the boxes fell off.'

'Well, I should think that's woken your sister if nothing else has.'

'Sorry, Mum,' he said again.

She ruffled his hair. 'Clumsy.'

Caroline left the room and crossed the landing to where Michelle slept. She peered into the room and saw that the girl was still motionless. The noise hadn't woken her. Unusual, thought Caroline, the girl was a light sleeper. She crossed to the bed with its poster of Boy George over the top and looked down at her daughter.

The girl was the colour of sour cream. Her lips were slightly parted and Caroline could hear low guttural breathing. Michelle's head was sheathed in a fine film of perspiration, her long brown hair plastered across her face.

'Michelle,' Caroline whispered.

The child didn't stir.

'Michelle.'

No response.

Caroline leant closer and shook her daughter gently. 'Wake up, love,' she said, her voice slightly louder.

Still silence.

'Oh God,' Caroline murmured. The girl felt so limp in her arms. 'Michelle, wake up, darling. Wake up.'

The girl stirred, sat bolt upright and opened her mouth to scream. Only a hoarse croak escaped. She saw her mother and grabbed her, pulling her close.

'Are you all right?' asked Caroline, feeling the perspiration on her daughter's body.

No answer.

'Michelle.'

The girl tried to speak but couldn't. All she was able to do was prod helplessly at her lips and tongue then her throat. She shook her head.

'Can you speak?'

She shook her head, tears welling up in her eye corners.

'Let me look at your throat,' Caroline asked and peered into the girl's mouth. There didn't appear to be any redness which would signal infection. She called Julie and asked her to get the doctor. Then, Caroline sat with her daughter, holding her hot body as if she were a baby.

Doctor Clive Fenwick arrived at 9.32 a.m. He parked his green Fiat outside the pub and knocked on the front door, his small black case in his left hand. As he waited for the door to be opened he brushed some spots of light rain from his jacket and smoothed down his thick black hair.

The door was opened by Julie. She ushered him inside and upstairs to the bedroom where Michelle was. David rose to greet him, the two men shaking hands warmly.

Fenwick frowned as he saw David's rash, the redness still prominent on his hands and face.

Pleasantries were exchanged briefly, then Fenwick set down his case.

'What's the problem, then?' he asked.

Caroline explained about Michelle. The child was sitting up in bed, her face still pale, her eyes still wide and dark-ringed. Fenwick sat on the edge of the bed and, while Caroline explained more details, he took the little girl's pulse and blood-pressure. Then he took the opthalmoscope from his case and looked into her eyes. Next he retrieved a wooden spatula from the case and asked Michelle to open her mouth. He depressed her tongue and shone a penlight on the back of her throat.

'Say aahh,' Fenwick instructed.

Michelle could only utter a low croaking rasp.

He frowned.

'Again.'

The same grating whisper came forth.

He sat still for a moment then took her temperature.

'Any idea what it might be?' asked Caroline.

Fenwick shook his head slowly.

'No idea at all?' snapped David and the doctor looked at him, taken aback by the tone of his voice.

Fenwick took the thermometer from Michelle's mouth and glanced at it.

'Her temperature's up. It's just over 101,' he announced. 'How long has she been like this?'

David and Caroline exchanged long looks, an unspoken word passing between them. Both realized that they must not mention anything that had gone on during the past two days and nights.

'Since last night,' Caroline stuttered. 'No, this morning, sorry. She was all right last night.'

'So, what could it be?' David wanted to know.

Fenwick shrugged. 'It looks like laryngitis,' he said. 'But it isn't.'

'What the hell sort of answer is that?' David snapped.

'She's running a high temperature, she's lost her voice but there's no infection in the throat. It's as if she's been frightened by something. People sometimes temporarily lose one of their faculties such as their voice, even their sight, after a powerful shock. Has anything happened that might have caused that?'

'No,' David said, quickly.

Caroline swallowed hard and clenched her fists at her sides.

'What can you give her, doctor?' she asked.

'Well, I can write a prescription for something for her throat and it might do the trick,' Fenwick said.

'Isn't *"might"* a bit vague,' David rasped.

Fenwick looked at him. 'You should let me have a look at that rash, David,' he said. 'How long have you had it?'

'Two days. It's an allergic reaction to something.'

'I'd like to examine you if you don't mind.'

'Is it necessary?'

'It can't hurt, love,' Caroline offered.

David shot her an acid glance and left the room.

'I apologise, doctor, I don't know what's wrong with him,' she said. 'In the last couple of days, his entire personality seems to have changed.'

'I noticed the difference,' said Fenwick. 'What about the rest of you? Anything wrong?'

'We're just tired,' Caroline told him.

He wrote the prescription for Michelle and handed it to Caroline, then he walked out onto the landing, surprised to find David standing there.

'If you want to examine me, fair enough,' he said. 'But make it quick. I've a lot to do before we open.'

Fenwick nodded and followed David into the next bedroom. He took off his shirt as instructed and Fenwick noticed that the rash only covered parts of his skin unprotected by the material.

'This is no allergy, David,' he said, looking at the red bumps which had risen on the pub owner's face, neck and arms. 'Do you have any pain?'

David shook his head. 'Not from the rash,' he said. 'I get headaches sometimes. Blurred vision. Nothing that doesn't pass.'

'And you've been like this for two days?'

'Yes.'

'Have you experienced anything like this before?'

'No.' But then again, David thought to himself, I've never been visited by a bloody UFO before. He almost managed a crooked smile.

'Caroline said you'd been moody, irritable.'

David snorted. 'Have you noticed any difference then?' he asked.

'Frankly, yes,' Fenwick told him. 'You look as if you could do with a few days' rest.'

David laughed bitterly. 'I've got a business to run. It's not as easy as that.'

'Isn't your health more important?'

David pulled his shirt back on. 'You *had* finished?' he said sardonically, and walked out. Caroline entered the room a moment later. She heard her husband's heavy footfalls on the stairs as he descended.

'Have you any idea what's wrong with him?' she asked, anxiously.

Fenwick smiled humourlessly. 'I'm afraid I don't seem to have been much help to you, Caroline,' he said apologetically. 'I can't be sure what's wrong with Michelle and I don't know what's the matter with David, but I know it isn't an allergy.' He paused. 'Has the rash spread at all, do you know?'

Caroline shook her head. 'I don't think so. It's not that that's worrying me, it's his moods and attitude. He blows up at the slightest little thing and . . .' She was about to mention the incident the previous night. She wondered if it would make her feel better to tell someone about what had happened to them, but she restrained herself. They had all sworn not to breathe a word to anyone outside the family other than the police, although for what good they were, they might as well not have bothered seeking their help. Caroline exhaled deeply. 'I just don't know what to do.'

'Is he worried about anything?' asked Fenwick.

She clenched her teeth together and shook her head. 'No,' she lied.

'Give it a couple of weeks,' he said. 'If he's no better, call me again. I'm just sorry I couldn't be of more help.'

She smiled thinly at him. 'And Michelle?' she asked.

'She'll be OK.' He squeezed Caroline's hand. 'Don't worry.'

She sought re-assurance in his smile but found none.

'You could do with getting away from here for a few days,' said Fenwick as they descended the stairs.

Caroline nodded. She showed him to the door where he said goodbye to David who was behind the bar. The pub owner nodded perfunctorily as the doctor left.

'Thank you,' said Caroline as Fenwick walked across to his car.

'Call me if there's any change,' he said. He climbed behind the steering wheel and started his engine. The Fiat disappeared down the driveway. As Caroline watched it go, she felt a sense of utter helplessness descending slowly around her.

It had been a hectic night in the pub and David was relieved when closing time came. His headache had got steadily worse as the evening had progressed until it felt as if someone were hitting him repeatedly with a red hot hammer. Now he stood in the bar drying the last of the glasses, pausing every so often to lean against the counter. Caroline was upstairs checking on Michelle and she returned a moment later with the news that the child was sleeping peacefully.

'Why don't you go and sit down, David?' she said, studying his haggard visage. 'Julie and I can finish doing this.'

He paused for a moment, then nodded almost imperceptibly.

'I could do with a breath of fresh air,' he said. 'I think I'll go for a walk.'

He glanced out of the window and saw that the rain had stopped, so he didn't bother to fetch his jacket. As he opened the door, King came bounding across the bar to join him and he allowed the dog to precede him out into the night.

The night was humid, probably, David thought, due to the light rain which had been falling for most of the day. The sky was mottled with cloud and he wondered if they were in for another storm. King bounded ahead, towards the hills which sloped up about two hundred yards to the left of the pub. Beyond those hills lay the woods where he had discovered the burnt patch. David did not feel like straying in that direction so he whistled the dog and went in the opposite direction. That way the ground was relatively flat, the low undulations which formed the edge of the hills flattening out into fields. About two miles beyond lay a farm.

David sucked in deep breaths, filling his lungs with the moist air and, out in the open, his headache began to diminish somewhat. He turned and looked back at the pub, lights burning in two of the downstairs rooms. It seemed a hundred miles away, receding into ever deepening blackness the further he walked.

He bent and picked up a stick, throwing it for King to chase. The dog hurtled off after it, finding it easily, despite the darkness and long grass. Returning it to David, the animal jumped about frenziedly until he threw it again. Once more King sprang off in pursuit.

David looked up at the sky and, for long seconds, was rooted to the spot.

A white light was emerging from behind a bank of cloud.

It was a second or two before he realized that it was the moon. David breathed an audible sigh of relief. Just lately, lights in the sky had come to mean something far more ominous.

He stood still, on top of a slight rise in the ground and looked over the land towards the farm which lay beyond. But for the odd light, it was in darkness. David wondered if anyone there had ever seen anything strange, experienced any of the things which he and his family had experienced? He doubted it. They would have reported it otherwise. Wouldn't they? His thoughts were interrupted by King's return. The dog jumped up excitedly and allowed David to take the stick from his mouth. He hurled it once more and the Alsatian disappeared after the piece of wood.

David heard a whimper then a loud snarl from the darkness.

He stopped walking, narrowing his eyes in an attempt to see into the gloom.

The growling became barking.

'King,' he called.

The dog did not return.

More barking.

'King. Come on, boy.'

The barks subsided into growls once more, low, angry sounds which grew louder. David knew that the dog must be close and he trod carefully now. Perhaps he had found a rabbit hole or something. He exhaled deeply. That was unlikely. The moon had disappeared, momentarily, behind a thick bank of cloud so the fields were bathed in darkness. David could hardly see more than ten feet ahead of him. But, finally, he caught sight of King.

The animal was standing perfectly still, shoulders up, the hackles at the back of his neck raised, and he was staring at something just ahead of him. As yet, David couldn't see what it was.

He got within two feet of King and the animal rounded on him. Teeth bared in a maddened grimace, he faced him and David wondered if the animal was going to attack him. King certainly seemed frightened, and David remembered that he'd seen the dog like this before.

The moon emerged from behind the clouds once again, throwing a cold white light across the land.

In this newly present brightness, David saw something just ahead of him and he felt a chill run up his spine.

The grass had been flattened in several places, and there were perhaps a dozen of the footprints the like of which he had seen in the wood just twenty-four hours earlier. He knelt down, the growling of the dog loud in his ears. King was facing the footprints again, teeth bared.

David inspected the marks from a safe distance. They looked identical. Each was about twelve or fifteen inches long, as before. But this time they were in an irregular pattern, as if the owner had been walking around aimlessly within a small area. And the other thing which puzzled David was their location.

They were in the middle of a field with no visible signs that they extended further. There was no sign of any landing by the craft. No scorched earth this time.

He took a few paces back, the dog also backing off.

There were more prints about twenty feet to his right, now clearly visible in the glow of the moon.

But there were no connecting steps between the two sets of prints. It was as if the creature had jumped from one spot to the other. But jumped twenty feet?

David looked around for more footprints but could see none. He checked his watch.

11.36.

But it couldn't be, he'd left the pub at 11.30. He'd been out for more than six minutes. David swallowed hard and lifted the timepiece to his ear. There was no sound from it. The hands were frozen and unmoving.

David, almost instinctively, looked up and cast a wary eye over the heavens. King was growling once again and David found himself back-tracking, his eyes straying once more to the footprints. He wondered if they were fresh. And if they were, where the creator of them was now. The thought made him shudder and despite himself, he turned and quickened his pace. He felt terribly afraid, exposed in this vast field. The pub was blotted out from view by the low hill, and he felt as if he were the only thing moving for miles. Visible to anyone who might be watching.

Anyone or anything . . .

He pushed the thought to the back of his mind, but nevertheless speeded up, breaking into a slow run. The dog scampered along beside him, barking intermittently, ears pricked as if trying to pick up some so far unheard sound.

Had the craft landed somewhere, he wondered? There seemed little else to explain the footprints. But if so, where? He felt his heart begin to beat faster. King ran ahead of him and now David broke into a full run, summoning up reserves of strength which he didn't know he had as the hill drew closer. Once over it he would soon be back inside the pub and, hopefully, safety. But, for the moment, all he could concentrate on was coaxing more speed from his churning legs.

He looked up as he reached the top of the hill.

The yellow egg-shaped object appeared as if from nowhere. David grunted something and ran faster, his eyes now on the glowing orb. He tripped and fell on the wet grass, rolling over and over until he reached the bottom of the short slope. He dragged himself upright, the pub now just a few hundred yards away.

High above, the brilliant yellow craft didn't seem to be moving, but then it dipped lower in the sky and began gliding slowly through the air.

King was barking wildly now and had already reached the front door of the pub. David saw the door opened by Julie who looked up to see him running towards her. He motioned her back inside, looking up again at the craft which was now over the pub and apparently motionless.

David reached the door, crashed inside and slammed it behind him.

'It's there again,' he said, breathlessly, his face and clothes stained with mud.

'What is?' asked Julie, vaguely.

'The bloody spacecraft,' he hissed. He crossed to the window where he was joined by the two women who scanned the dark sky for any sign of the fluorescent ovoid. Caroline was the first to spot it.

'I found more footprints,' David said. 'In the fields over there.' He motioned towards the direction from which he'd come.

The craft seemed to be holding its position. It didn't, as it had in the past, dive lower. It merely moved back and forth in the sky like a bead being pushed along an abacus. David looked round to see that the clock above the bar had stopped, and when he asked the two women, both found that their watches too had ceased to function.

'Why haven't the fuses blown?' asked Caroline, softly.

'I don't know,' David said. 'Perhaps it's too high up to affect the circuits.'

The craft disappeared out of sight, only to return a moment later. It moved relatively slowly, its glow much less dazzling than usual. David assumed that was because it was so high up. As they watched, it vanished once more, only to return moments later. David frowned and wandered through into the sitting room, peering out of the bay window.

He saw the craft from where he stood. Saw it glide away behind a bank of cloud. By the time he reached the bar, it was there again.

'It's circling us,' he said, flatly.

Caroline looked puzzled.

'It's circling the pub,' David repeated.

All three of them watched it appear then disappear as it continued its unexplained manoeuvre.

'What is it doing that for?' Julie wanted to know.

'Perhaps it wants to make sure we're still here,' said David, sardonically.

Silence descended as they watched.

For a full hour the craft circled then finally, it disappeared.

This strange ritual was to continue for several more nights.

Ten

August 6th

The banging was loud.

David rolled over in bed, the sound disturbing him, yet he still clung to sleep.

The banging came again, more insistent this time.

He opened his eyes, at least he thought he did. Was the noise part of a dream from which he had yet to awake? He heard it once more. A steady pounding rhythm, then a few moment's silence. Beside him, Caroline stirred also but did not wake up.

More banging.

Jesus, what the hell was going on? He sat up quickly and looked at the bedside clock.

7.04 a.m. The alarm hadn't even gone off yet.

The banging was coming from downstairs. Someone was knocking on the front door. David swung himself out of bed, running a hand through his hair, rubbing his eyes in an effort to wake himself up. He'd better get downstairs and open the door before whoever it was woke the entire family. It wasn't a drayman, he was sure of that. For one thing they always used the rear entrance and, for another, he wasn't expecting any deliveries for a day or two. Whoever was making the racket was banging on the front door. But who, he wondered, would be knocking on the pub door at such an early hour in the morning?

Perhaps someone was lost and needed directions, he mused as he descended the stairs and headed through into the bar.

More banging.

All right, David thought to himself, don't knock the bloody thing down, I'm here now. He slipped back the bolts which secured the door then unlocked it, pulling his dressing gown around him as he pulled the door open.

Standing on the top step were two men.

Both of them were dressed in black suits, complemented by brilliant white shirts. Black ties, shoes and hats completed their attire and they reminded David of undertakers. The only things other than their shirts which weren't black, he noticed, were their gloves. Each man wore a pair of what looked like grey suede gloves. Beyond them, parked in the driveway, was a gleaming black Mercedes.

'Mr Ellis,' said the first of the men, and it sounded more like a statement.

David ran an appraising eye over both of them. They looked like twins, a fact enhanced by the identical clothes which they wore. Both men were in their fifties, a little over six feet tall, David guessed, and both were deathly pale.

'David Ellis,' the second one said.

'Yes,' he answered. 'Who are you?'

'If we might come in,' said the first.

For some reason, David stepped back and allowed the men into the bar, closing the door behind them. For brief seconds he felt his head spin, much as it had that night out in the woods, but the feeling soon passed and he faced the two curious visitors.

'I asked who you were,' said David, studying the men. They stood perfectly still, arms by their sides, eyes fixed on David, their gaze unswerving. So intense in fact that it made him feel uncomfortable.

'Where is your wife?' asked the second one.

'Look, what the hell is this?' demanded David. 'Now, for the last time, who are you and what do you want here?'

The first man coughed lightly, raising one gloved hand to his mouth. When he removed it, David saw something smeared on the material. Something which looked like lipstick. He frowned and looked at the two men more closely.

Both, by this time, had removed their hats and David saw that they were completely bald. Not only that, but neither man had eyelashes or eyebrows.

David regarded them warily.

'If we might sit down, Mr Ellis,' said the second man.

Again, almost without hesitation, David led them through into the sitting room where both men sat down on the sofa.

There was a long silence during which the two men seemed not to be looking *at* David but *through* him, as if he didn't exist.

'When was the first time you spotted the spacecraft?' asked the first one.

David's mouth dropped open.

'How did your children first come to find the footprints in the wood?' the second man added.

David felt as if his jaw had locked. He wanted to speak but couldn't.

'Did your wife and her sister obtain a good look at the alien which came from the spacecraft?'

'Did the doctor manage to help your daughter?'

The questions flew back and forth, asked alternately. David could only sit dumbfounded as the two men continued.

'What do you want to tell us about what you have seen, Mr Ellis?'

David finally seemed to come out of his trance-like state, but, when he did speak, his voice was low. Tinged with fear, rather than anger. Who the hell were these men? How could they know what had happened? They had told him things which no one else could possibly know. He swallowed hard, aware of the fact that his heart was beating just that little bit faster.

'How do you know all these things?' he asked. 'Who are you?'

'Is it important, Mr Ellis?' asked the first man.

'Yes, it is.'

'Who have you told about what has happened here?' the second man wanted to know. 'Anyone other than the police?'

David eyed the two men warily. He glanced out of his eye corner towards the kitchen door. Inside the room, King and Sabre slept. David wondered why they had not barked when they first heard the door being knocked. Yet, even now, with voices so close to them, they remained silent.

'The police did not believe you, did they?' said the first man.

David shook his head.

'We all agreed not to mention the incidents to anyone.' He found that the words came easily.

'That is a good thing,' the first man told him. 'You should not talk about what has happened to anyone. Do you understand?'

All David could do was nod. Once more he felt that peculiar light-headed feeling come upon him, and when he tried to move, his arms and legs felt heavy.

The first man coughed again, shaking David from his dream-like state.

'You still haven't answered my question,' he said. 'Who are you?'

'That does not matter.'

'It matters to me,' David snarled.

'We are concerned in matters such as this,' the second man told him.

'But how do you know what has happened here?' David demanded, raising his voice.

'We know. That is all that is important.' The second man said.

'How did you find the pub?' David wanted to know.

Even if the men had intended answering, they were prevented from doing so by the arrival of Caroline. She stood in the sitting room doorway looking at the two black-clad men in bewilderment. Despite the fact that both of them had their backs to her, they turned round and looked at her, as if sensing her presence.

'Mrs Ellis,' said the first man.

'I heard voices,' she said to David.

'We have been speaking to your husband,' the second man

told her. 'We also wished to speak to you.'

'About what?' she enquired.

'They know about what's been happening here,' David said. 'The UFOs, the aliens, everything.'

'How do you know?' Caroline asked them, looking closely at the men. She too noticed that they were completely devoid of facial hair. She also noticed how red and shiny their lips were. They seemed to give off a strange odour, rather like old make-up. The smell usually associated with a theatre. Their suits were immaculate, the creases in the trousers razor sharp, not a speck of dust or fluff to be seen. They looked as if they had both just walked out of a dry-cleaners.

'It is best that you say nothing. You are also to say nothing about this visit. Do you understand?' the first man said.

Caroline nodded slowly, gazing into the man's eyes. He seemed to have no pupils, merely large dark irises. She felt as if she were drowning in those large black orbs.

Both men got to their feet as if at a given signal.

'You're not leaving here until I find out who the bloody hell you are,' David snarled.

'We are leaving now, Mr Ellis,' the second man told him. 'Do not try to stop us.'

David clenched his fists but he could only stand and watch as the two black suited men walked towards the front door. He and Caroline followed them, pausing a few feet away.

'You have two rings in your pocket, Mrs Ellis,' said the first man, pointing at Caroline's dressing gown.

'Yes I do,' she said, dumbfounded. Had he heard them chink together when she got up?

'Give one to me.'

'Why?' she asked.

'I asked you to give one to me,' he repeated.

She did as instructed, laying a gold signet ring on the outstretched palm of the first black suited man. He closed his gloved hand around it, his arm held out stiffly before him.

'Do not tell anyone of this meeting,' he said.

'Are you threatening us?' David said. 'I'll have the police out here if I see you again.'

'Do not tell anyone of this meeting,' the second man echoed. Then he turned as his companion opened his hand.

Caroline's ring was gone.

The first man lowered his hand and opened the front door.

Both David and Caroline stood watching as the two men made their way across to the black Mercedes and, only then, did David notice something odd about it.

There was no number plate. No lights. No indicators.

They watched the car drive away, disappearing down the long driveway and, for a moment, David considered climbing into the Volvo and following, but his thoughts were interrupted by a startled cry from Caroline.

The gold signet ring was back on her finger, there in plain sight as she stood with her hand outstretched.

David and Caroline sat in the kitchen over a pot of tea, neither of them able or willing to speak. There didn't seem much to be said. Every now and then she would glance at her signet ring and touch it tentatively, looking up at David as if seeking an answer from him.

'I wish to God I knew who they were,' he said, watching the steam rise from his tea cup.

'They didn't tell you anything about themselves?' Caroline asked.

He shook his head.

'How could they know what has been happening here?' He paused. 'Unless . . .' He allowed the sentence to trail off. 'That night the spacecraft landed and that . . . thing got out. What did you say it looked like?'

'It's difficult to describe,' Caroline said. 'I don't see what that has to do with those two men.'

'Did it look like them?'

She shook her head. 'It was much bigger.'

David nodded slowly, resignedly. Beside them, the dogs still slept.

'Could those men have been from the Government?' asked Caroline. 'Investigators or something like that?'

David shook his head.

'I doubt it. Besides, why would they need to be so secretive and mysterious? They could have told us straight out.'

'Perhaps they thought if we told other people it'd cause a panic.'

'They weren't from the Government.'

'How can you be sure, David?'

'If they were, how do you explain the trick with the ring?' he said, challengingly.

Caroline twisted the gold band gently. It felt slightly warm to the touch and had done since the black suited man had returned it.

'What are we going to tell Julie and the kids?' she asked.

'Nothing,' he said, flatly. 'They've been through enough already.'

Caroline nodded then looked down at her hand. She continued toying with the signet ring, feeling the strange heat which ran through it. She wondered if they would see the men in black again. The thought made her shudder.

Eleven

August 10th

David stood before the bathroom mirror and peered at the image which stared back at him. He tentatively touched his skin, both surprised and relieved when he felt no pain or irritation. The rash had all but disappeared, the redness had gone. Only the peppering of small bumps remained, and they too were fading.

That morning, for the first time since the incidents had begun, he had awoken without a headache. He felt brighter and, he told himself, he looked better.

Michelle too had begun to progress. Although she was still unable to speak, her temperature had fallen and she had regained her customary colour. They had agreed that if her voice hadn't come back within the week, they would have to ask Doctor Fenwick to come back, or take Michelle to hospital. The previous day she had been out on the hills with Dean and the two dogs.

David still forbade them to go near the wood where the spacecraft had landed but both the children seemed quite happy to comply with this arrangement.

There had been no sign of the glowing light for four days now and, in fact, not since the appearance of the mysterious men in black had anything unusual happened. Nevertheless, David crossed from the mirror to the bathroom window and opened it, peering out into the clear night sky. It was clear but for a smudge of cloud here and there. The air smelt fresh and David afforded himself a smile as he flicked off the bathroom light and wandered into the bedroom.

Caroline was sitting up in bed combing her hair. She smiled at David as he climbed in beside her and was happy to see the gesture returned.

She had noticed a change in him during the past four days. A change for the better. Slowly, but surely, he was being transformed back into the David she knew. The man he had been before all these crazy things started to happen. His moods had lessened and he wasn't so irritable. He was still prone to the occasional outburst, but his reactions were nowhere near to being as extreme as they had been.

She finished combing her hair and laid the comb on the bedside table, then she settled herself on the bed. David flicked off the light, leaning across to kiss her gently on the lips as he did so. It was the first time he had done so for over a week.

It was warm in the room, despite the breeze which blew in through the open windows. The curtains billowed like thick mist and Caroline watched them, her eyelids gradually becoming heavier as she drifted off into welcome oblivion.

Julie, in the room across the landing, stood gazing out of the window, eyes moving furtively back and forth across the sky as if searching for something which she hoped she wouldn't find. She too saw only the odd patches of cloud. After fifteen minutes or so she returned to bed, pulling back the covers until only a sheet covered her. The heat inside the room seemed to be increasing, yet still a steady and constant breeze wafted in. She was hot-blooded, she told herself and smiled thinly. That was what her husband had said. Almost unconsciously her hand touched the gold band on her wedding finger. Despite having been separated for over three months from her husband, she still wore his ring. A vision of him flicked into her mind but she brushed it aside. It *was* getting hotter in the bedroom, she was certain of it.

She eventually drifted off to sleep.

No one inside the pub heard the low buzzing sound which cut through the air so insidiously, gradually growing louder.

Despite the increased warmth in the building, no one stirred.

At exactly 2.19 a.m. every clock and watch in the pub stopped.

High above, in the cloudless sky, a light, too large and too bright to be a star, hovered. An egg-shaped yellow light.

The craft remained motionless for a moment, then began circling.

Caroline felt as if she were floating.

She put her arms down by her sides but there was nothing there to support her. It was as if she were suspended on a cushion of air like some kind of human hover-craft. She tried to open her eyes but couldn't. Even though she couldn't see herself she knew she was naked. A gentle, soothing heat seemed to caress every inch of her body, back and front. She was lying flat, but on what she didn't know. Her arms felt like lead weights hanging beside her. She tried once more to open her eyes and this time succeeded. For a second she wished she hadn't. Blinding white light caused her to gasp aloud. No sound came forth when she opened her mouth. The light was all she could see and she winced at its severity. She tried to move her head but found the effort impossible. Caroline was held, by some kind of invisible clamp, as securely as if she had been manacled. She couldn't even raise her head, only by looking down could she see for sure that she was naked. And when she managed to swivel her eyes down she felt a sharp pain at the back of her head. She contented herself with gazing straight up into the dazzling brightness. It was like staring directly into an arc light.

Something touched her left ankle and she tried to jerk her head up, alarmed by the sensation. With mounting panic, she found that she was unable to see.

She felt something brush against her other ankle.

Hands?

They closed over the joints and she felt her legs being parted.

She tried to scream but no sound would come.

The hands on her ankles felt cold, like dead fish. Her body

remained perfectly still as her legs were forced yet further apart. But she felt no pain, just growing fear and an overwhelming sense of helplessness. She tried to raise her arms, her head. Caroline could not move and now the hands were moving up to her knees, forcing her legs to their widest limit. God help me she shrieked, but the sound was inside her head. She could hear nothing and feel only the cold deathly touch of those unseen hands on her legs.

Another cold hand gripped her wrist and began to raise her left arm until it was level with the rest of her body. The same thing happened to her right arm.

Spreadeagled, she was suspended by some unknown and invisible power, held by hands that felt as icy cold as the hands of a corpse.

But suddenly there was a feeling of unbearable heat. It began around her navel, spread downwards to her pelvis and thighs then upward to her breasts and neck.

It was at that point she screamed again and, this time, the sound was not just inside her head.

David was hurled from sleep by not one but two loud shrieks of terror.

He sat up to find Caroline shaking uncontrollably, her eyes bulging madly. He gripped her and pulled her close.

'What's wrong?' he asked.

She tried to speak but couldn't. Her lips moved noiselessly like some kind of mime artist.

'Caroline.' He shook her.

She was breathing rapidly, her exhalations coming out as swift racking sobs.

'Calm down,' he said.

She stopped shaking slightly and managed to swallow.

'What's wrong?'

'A nightmare,' she gasped. 'I had a nightmare.'

He laid her back on the pillow, holding her hand, feeling how hard her heart was pounding as she held his comforting hand to her chest.

'Oh God, David, it was terrible.'

He brushed a hair from her face. She suddenly jerked her head around to face him then swiftly hauled herself out of bed.

'What are you doing?' he asked, following her.

'Julie,' she said and dashed out onto the landing.

Her sister was already there, her face pale and ghostly, her eyes wide and staring.

'I had the most terrible nightmare,' Julie said.

'So did I,' Caroline confessed.

'I couldn't see, everything was too bright. The light was blinding.'

Caroline's face froze in an expression of horrified recognition.

'I felt as if I was floating,' Julie continued. 'But I couldn't see anything except the light and then I couldn't even open my eyes.'

'Did it feel as if something was holding your legs and arms?' said Caroline.

Julie nodded, a look of surprise on her face.

'Yes, but . . .'

'Someone forcing your legs open.' It came as a statement rather than a question.

'I was naked, too, but Caroline, how do you know?'

'Did you feel a burning pain here?' She pointed to the area around her own navel.

Julie nodded, still bewildered.

'I had the same dream,' Caroline told her.

'But it was so vivid. I felt helpless. I couldn't move, I couldn't speak.'

Caroline nodded in affirmation.

David appeared at her side. 'What's going on?' he asked, looking at the two women.

'That nightmare I had,' Caroline said. 'Julie experienced one exactly the same.'

David frowned. 'Exactly?' he asked, somewhat sceptically.

'Everything about it was identical,' Julie assured him.

'Well there's nothing too sinister in that is there?' he said.

'But it is a little odd, isn't it?' said Caroline.

David smiled humourlessly. 'Most of what's been happening around here lately has been odd,' he said, sardonically.

'Are you all right, Julie?' Caroline asked her.

Her younger sister nodded and smiled weakly. Then she stepped back inside her bedroom. They exchanged goodnights and returned to their own room, settling themselves in bed once more.

'What exactly did you see?' David asked.

Caroline told him. 'And Julie experienced the same thing. Down to the last detail.'

There was a long silence and, in that silence, David noticed that the clock on the bedside table had stopped, as had his watch. Caroline saw that hers too was no longer working. Immediately, David swung himself out of bed and hurried across to the window, half-frightened to look for fear of what he might see. He studied the scudding clouds, watching for any sign of movement or the slightest suggestion of light other than that cast by the moon. He stood there rigidly for five or six minutes then returned to bed.

'Nothing,' he said, quietly.

'David, the light that I saw in my dream, it was like . . .' She allowed the sentence to trail off.

'Like the light from the spacecraft?' he said.

She nodded. 'It was as if I was inside it.'

He gripped her hand and noticed how cold it was, especially around the wrist.

They settled down and David turned over. He was asleep within minutes. But Caroline merely lay there gazing into the gloom, hardly daring to close her eyes for fear that the nightmare would return once she gave herself up to sleep. She huddled closer to David and continued watching the bedroom curtains as they billowed wildly in the breeze.

Across the landing in her own bedroom, Julie did not immediately return to bed. She flicked on her bedside light and sat down in front of her dressing table, gazing at her reflection in the mirror. Her eyes looked bloodshot and watery.

As if she had been looking into bright light?

She felt cold, her skin rising into goose-pimples as she sat there. Julie rubbed her arms and, as she did so, she noticed something. Looking more closely she saw that there were several white marks on her wrists, like indentations.

She shuddered and extended her legs.

The marks were on her ankles too.

Twelve

August 11th

David stood to one side as the two burly draymen rolled the barrels of beer down into the cellar by way of a makeshift gantry. They always called early and today was no exception. It was only just 7 a.m., but the fresh morning air made him feel pleasantly alive and high above the sun was already cruising its way into the cloudless blue sky. David helped with the last of the kegs, then dutifully signed the delivery form. He exchanged pleasantries with the two men, both of whom had been delivering to him for the past six years. They chatted about football, they chatted about pubs in general, one of them even mentioned the weather, then they made their way back up from the subterranean depths to the lorry and drove off to make their next delivery. David asked them to leave the cellar bulkhead open to blow away some of the mustiness in the air.

He rolled up his shirt sleeves and began moving the kegs to their appointed positions.

It was a large cellar, well stocked with everything from beer and spirits to some fine old wines. The Horseshoes might be off the beaten track but, David thought, that was no reason to limit what it could offer to those fortunate enough to find it.

He passed the fuse box and opened it to check on the fuses inside. They had not blown last night, nor for the previous four nights. He closed the box again, as if fearing that prolonged perusal of the objects would cause them to short out at any moment.

He stood still when he heard a loud buzzing sound.

It was close by him and he spun round, eyes anxiously searching the underground for the source of the noise. He could see nothing but the buzzing was still present.

A wasp flew past him.

David smiled thinly. Watch the paranoia, he told himself.

The wasp was followed by another, then another and the pub owner frowned as he saw them disappear into a corner of the cellar. A corner untouched by the rays of sunlight streaming in through the open bulkhead doors. He stopped what he was doing and wandered across to the spot where the wasps had disappeared.

'Christ,' he murmured, taking a step back.

The nest was as big as one of the two beer barrels which flanked it. Conical in shape it seemed to hang there and David felt a sudden ripple of fear run through him. However, apart from the two or three insects moving about on the surface of the nest, the nest appeared to be quiet. He ran appraising eyes over it once more, marvelling at its size. God knows how many wasps it held. He had read somewhere that the average nest was about 20,000 strong, but this one looked as if it might hold four or five times that number of the striped creatures.

David took another pace back, his eyes glued to the wasps in view as if he were convinced they were watching *him*. He decided the best thing to do was to get in touch with the local Rentokil man and have him remove the bloody thing before the insects decided to swarm or something like that. He turned and made his way up the set of stone steps which led to the kitchen.

Caroline was cooking breakfast. She always got up with him on delivery days and, particularly after her dream the night before, she felt as if she would rather be up and about than lying in bed. She hadn't slept much for the remainder of the night and it showed in the dark rings beneath her eyes. David returned from the sitting room clutching a copy of the Yellow Pages.

'What are you looking for?' asked Caroline.

He told her about the wasps' nest in the cellar.

Caroline turned away from the pan of bacon and eggs for a moment, a troubled look on her face.

'Isn't that dangerous?' she said.

'That's why I'm calling the Rentokil bloke,' David explained. 'It should be OK as long as it's not disturbed in the meantime.'

He ran his finger down the list of names on the selected page, murmuring to himself as he read them off.

The phone rang.

David looked up. It was a bit early for anyone to be calling. He watched Caroline as she crossed to the phone and picked it up.

'Horseshoes public house,' she said, gaily.

There was no sound from the other end.

'Hello,' she said. 'This is the Horseshoes public house. Can I help you?'

'You will tell no one about what happened to you last night Mrs Ellis.'

The voice, although sounding somewhat metallic over the phone, was clear enough and she shuddered. It had an uncomfortably familiar ring about it.

'Who is this?' she said.

David got up and crossed to the phone, seeing the look of concern on her face.

'What is it?' he whispered, pointing at the receiver which Caroline was gripping so tightly her knuckles looked bloodless.

'You will also instruct your sister to keep silent,' the voice said.

'I don't know what you're talking about,' Caroline said, tilting the receiver to one side so that David could also hear the voice.

There was a moment's silence.

'What happened to you and your sister last night must remain secret. Do you understand?'

David raised his eyebrows in surprise.

'You have no right to give me orders,' Caroline said. 'Who

are you anyway?'

'You have knowledge of us,' the voice said, then the line went dead.

Caroline flicked the cradle but nothing happened. She couldn't even get a dial tone. There was a crackle and for a second she felt the phone growing hot in her hand. Was she imagining it? No. The plastic was getting warmer. She put it down hurriedly. After a moment or two she tentatively lifted the receiver. The continuous purring had returned, the phone seemed to be working again.

'What the hell was that about?' David asked.

Caroline looked shaken. She told him what the voice had said. About her dream. About Julie's dream.

'You know, that voice sounded familiar,' David said, racking his brains to try and visualize where he'd heard it.

'The men in black,' Caroline said, flatly. 'It sounded like one of those men who came here the other day.'

David clicked his fingers and nodded. She was right. But how could they know what had happened the previous night? The same way they had known about all the mysterious phenomena, he thought, answering his own question. David stroked his chin thoughtfully. The conversation had been worded in a peculiar fashion. The voice at the other end of the line had not referred to Caroline and Julie's nightmares but rather had mentioned the imagined goings-on as if they had been recognisable 'events' which had actually taken place. Yet he knew this to be impossible. Nevertheless, Caroline was right; the voice had most definitely sounded like one of the mysterious black clad visitors they'd received five days before. The tone, too, had been the same. A veiled warning to keep silent. And if they decided not to? David wondered what kind of retribution, if any, would be exacted against them, but he decided not to dwell on that particular eventuality.

Julie padded through into the kitchen and smiled at them. Caroline looked at David, wondering whether or not she should mention the phone call to her sister. As yet, only David and Caroline knew of the existence of the men in black; to tell Julie of the phone call would necessitate mentioning the funereal visitors too. But the caller had mentioned Julie. Caroline felt that she had a right to know.

'Julie,' she paused, looking at David. 'How do you feel this morning?'

Julie poured herself a cup of tea and sat down opposite David.

'Tired,' she said, looking down at her own wrists but not seeing the white indentations she'd spotted the night before. They seemed to have vanished.

'There was a phone call a moment ago,' Caroline began, noticing David out of her eye corner. He was glaring at her, almost shaking his head as if to prevent his wife from continuing. 'A man called, he mentioned . . .'

Caroline was cut short by a sudden and deafening eruption of noise. A loud buzzing which seemed to bore right through her skull. From the window, there was a dazzling flash of light and all three of them turned to see what was happening.

Hovering outside the pub, little more than ten feet off the ground, was the yellow, egg-shaped craft.

There were several loud bangs from below and David knew that the fuses had blown again.

'Get down,' he shouted and ducked beneath the table, pulling the two women with him. They crouched on the floor, eyes fixed on the shape outside the window which was now burning with a brilliant white light. The humming sound became almost unbearable, but now it was joined by a more strident, venomous buzzing from below them.

The glowing craft drifted higher but the subterranean buzzing seemed to grow louder and David, with mounting horror, realized what it was.

The door which connected the kitchen to the cellar was open.

He scrambled to his feet and tried to reach it but he was too late.

The first of the wasps came through the door, like the scout for a flying yellow and black army. A huge swarm of the stinging insects, so numerous they appeared to be one undulating floating mass, swept up from the cellar and into the kitchen. A thinner trail of them followed, like the tail of some monstrous tadpole.

Julie screamed, shouting in pain as the first of the insects stung her with its rapier like sting. She crawled away, looking at the frenzied horde in horror. They seemed to fill the kitchen, some of them now pouring through into the sitting room.

David yelped as he was stung, swatting at the insects with his hands as they flew around him.

The dogs seemed uncertain what to do. Dive-bombed by numerous wasps they snapped at the stinging creatures, also growling their anger and fear at the glowing spacecraft, which by this time had risen out of sight and was hovering just above the pub.

David looked out of the window to see yet more of the wasps rising, fountain like, from the bulkhead doors. They seemed to rise up like spouting oil before dispersing into the air.

Inside the building he, Caroline and Julie swatted the maddened things with any implements they could lay hands on. But the wasps seemed more intent on reaching the outside. The three humans were little more than obstacles in their path. David took the chance and hurled open the back door. He knew, even as his hand touched the wood, that one of two things was going to happen. Either the insects already inside the pub would fly out into the fresh air, or those outside would swarm in to add to the pandemonium. He prayed it wouldn't be the latter.

As the door swung open, both dogs too, ran for the outside and, David was relieved to see, so did the wasps, the bulk of them pouring out of the opening into the sunshine beyond.

But still the buzzing continued and he realized that it came from the spacecraft which was still above them. In the panic caused by the wasp attack, he had forgotten about the glowing ovoid, but now, as it dipped into view once more, he felt the hairs on the back of his neck rise stiffly. Slamming the door, not caring how many of the wasps he trapped inside the kitchen, David locked the partition against the greater malevolence of the intruders.

Many of the wasps were still flying around outside, formed into a colossal and ever-moving miasma of yellow and black, but their own frantic sound was drowned by the hum of the hovering craft.

It swung and dipped at an impossibly low angle, then as David watched, it began to rise, slowly at first then with breath-taking speed until finally it disappeared into the depths of the blue sky, like a stone sinking into the sea.

Now only the wasps remained but, as he watched, they seemed to turn as one and fly off across the garden at the back of the pub, heading for the hills and the woods beyond. In a matter of moments every last one of them had disappeared from view.

Inside the building no one spoke. The kitchen was wrecked. The table had been overturned and all three of them had been stung at least three times. A milk bottle lay smashed on the lino, its contents seeping across the floor.

'Oh God,' murmured Caroline, getting to her feet, wincing as she touched the places she'd been stung.

'Where did the wasps come from?' asked Julie, squatting dazed and shocked beside the sitting room door.

'There was a nest in the cellar,' David told her. 'Perhaps the frequency of sound that the spaceship was putting out disturbed them.'

He opened the back door and saw that the two dogs were standing on the lawn, heads tilted skyward. David too looked up but was thankful to see nothing more than the glinting sun. Daylight. It was the first time the craft had appeared in daylight. Why the new and unexpected boldness he wondered? He sank to his knees on the grass, sucking in huge lungfuls of air, trying to ignore the pain from his numerous stings. He hadn't known which to fear the most. The swarming, stinging wasps or the phosphorescent ovoid. Perhaps the wasps, he thought. The craft was all too familiar to them by now. However, never before had it appeared so close and in daylight too. He regained his breath and got to his feet again, wandering back into the pub. David made his way down to the cellar while Julie and Caroline attempted to put the kitchen back in some kind of reasonable order. It was cool in the underground cavern and the smell of damp wood and stale beer was as welcome as that of fresh air had been only minutes before. But there was another, more pungent odour to assault his nostrils. The smell of burnt out fuses.

The door of the fuse box had been blown right off, it lay, blackened and twisted, several feet away, and as David bent to pick it up he saw that it was still glowing red hot in places. He left it lying there and turned his attention to the fuses themselves. Or at least what was left of them.

It was as if each time the craft appeared, it sucked the power from the pub, almost as if it needed it. Like some kind of parasite it fed on the electricity, using it for its own ends. But, David thought, what would a craft as sophisticated as that need electricity for? Creatures able to build and fly a craft which could travel countless millions of miles would be using a fuel or power far beyond anything within man's comprehension. At first he had thought the spaceship was probably acting as some kind of storage machine, using up power, attracting it as surely as a magnet attracts iron filings. Only now it also seemed able to control animal life too. He looked around and saw that the wasps' nest was little more than shredded pieces of pulp. It looked as if the insects had torn their way out, so anxious were they to be free. He swallowed hard, thinking of the damage and injury which the nest could have inflicted.

After much effort David managed to repair the fuse box, replace the fuses, and was relieved to find that the supply was in proper working order once more. That done, he wandered upstairs again and across to the phone. There, he dialled Matlock police.

There was a moment's pause then the phone was picked up.

'Matlock Police station.'

'I'd like to make a report,' said David.

'About what, sir?' the voice at the other end wanted to know and David recognised it as Sergeant Gregory.

'I want to report a UFO.'

A moment's silence.

'Come again?'

'I said, I want to report a UFO.'

'Who is this?'

'My name's David Ellis, I own The Horseshoes pub about two miles outside the town.'

Gregory coughed.

'Ah, Mr Ellis, I remember,' his tone had taken on a weary sound. 'What is it this time?'

'I told you,' David repeated. 'I want to make a formal report of a UFO sighting.'

'Look, Mr Ellis, we've been through this with you before. I've sent men out there and you've never been able to prove anything. What do you want me to do?'

'I want a man out here now,' David snapped.

'Sorry,' said Gregory, quickly.

'What do you mean, sorry?'

'I mean, I'm not sending anyone. My men have got more important things to do than go chasing about looking for bloody lights in the sky.' The sergeant's tone softened for a moment. 'I wish I could help you, Mr Ellis but, well, this is the third time you've reported something like this and there's never been anything to back it up. I'm sorry, I can't send anyone.'

'You mean you *won't*,' snapped David.

Gregory's tone hardened. 'All right, if you want to put it that way. No, I won't. And, Mr Ellis, I'd appreciate it if you didn't bother me again about flying saucers, like I said, I have got more important things . . .'

The sentence was cut short as David slammed the receiver down. He leant against the wall, his breath coming in short, angry gasps.

'Bloody fools,' he rasped.

Caroline looked quizzically at him.

'The police,' he told her. 'They won't send anybody out here.'

'Why not?' she wanted to know.

'Because by now I should think they're convinced we're all stark raving mad.' He banged the wall angrily. 'Maybe they're right.'

'But David, someone's got to do something. We need help.'

'You try telling them that,' he said, pointing to the phone.

'What about getting in touch with a newspaper?' Julie suggested.

David spun round. 'You're kidding. That's the last thing we want. Newspaper reporters crawling all over the place. We swore we'd keep this to ourselves, as far as possible anyway. If a newspaper gets hold of this story there'll be people coming from miles around just to have a look at *us*. Come and see the Ellis family, the nutcases who keep on seeing UFOs,' he said, acidly. 'The papers would have a field day.'

'Then who is going to help us?' Caroline wanted to know.

David shrugged his shoulders helplessly.

'Perhaps it would be better if we just moved away from here,' Caroline suggested.

'No,' David said. 'Never. We're not going to let this beat us.'

'I don't care whether it beats us or not, I don't want it to kill us. Besides I'm frightened for the kids. We can't know what long term effects it's going to have on them. It's shaken *us* up badly enough so God alone knows what it's done to them.'

'Well, we're not moving. These things started happening suddenly, they could end just as suddenly.'

'And if they don't?'

'It can't go on forever,' David said, exasperatedly.

'I know that, I wish we knew *how* it was going to end, though,' Caroline said.

'I just can't understand the police,' said David. 'It's as if they don't want to get involved. I mean, there are thousands of UFOs reported every year, at least that's what you'd think looking in the papers. I mean, it's not as if we're asking them to do anything other than make a formal report. If we'd claimed to have seen visions of The Virgin Mary then I could understand them being sceptical, but . . .' He allowed the sentence to trail off.

'Do you think those men had anything to do with the police or the Government?' said Caroline, letting it slip quite innocently.

Immediately, David glared at her but his gaze rapidly shifted to Julie who looked on in bewilderment.

'What men?' she asked.

'You might as well tell her,' David announced, wearily, then he listened as Caroline relayed their encounter with the mysterious men in black. She also mentioned the phone call earlier that morning.

Julie sat down, stunned.

'And they didn't say who they were?' she added.

'No,' David told her. 'I asked them enough times but they wouldn't say. I just can't understand *why* the police won't help though.'

The three of them sat in silence for a long time, then David looked up at the two women.

'I know who might be able to help us,' he announced.

They looked at him questioningly.

'Roy Campbell. He's been interested in astronomy and bloody flying saucers for years. He's even got his own telescope set up in a shed in his back garden, it drives his wife mad. I think he's part of some kind of organisation that checks on things like this.'

'How do you know, David?' asked Caroline.

'He told me a while ago, even offered to show me the telescope.' A slight smile spread across David's face. He got to his feet and fetched the telephone directory, flipping through the pages until he found what he was looking for. He read out Campbell's number.

'Are you going to call him now?' asked Julie.

'Not about what's been happening,' David told her. 'I just want to check that he's coming out here today. When he does I'll have a quiet word with him.'

'He's out here nearly *every* day, David,' Caroline said.

'And anyway, how do you know you can trust him not to blab to someone else?'

'Roy's not that kind of bloke. People don't take him seriously anyway, not as far as his outer space talk goes. Even if he *does* tell anyone they won't believe him.' David picked up the phone book then crossed to the phone and dialled the number. It rang three times then he recognised Campbell's voice on the other end.

'Hello, Roy, it's David Ellis out at The Horseshoes.'

Campbell sounded pleasantly surprised and asked what David was calling for.

'I just wanted to see if you were coming out here later.'

Campbell said that he was.

'I'd like a word with you about something.'

Campbell wanted to know what it was.

'It'll wait until you get here.'

They exchanged goodbyes and David hung up. He returned to the table and sat down again. Upstairs, the children were moving about, they would be down soon.

'We'll see what Roy has to say about all this,' David said. 'He should be able to help us.'

'And if he can't?' said Caroline ominously.

A heavy silence descended.

Thirteen

The Horseshoes was relatively quiet for a lunch time, just the regulars sitting at the bar and five or six other people at the tables elsewhere. There were two young couples out in the garden at the back enjoying the sunshine which was now at a peak. David chatted with his customers, but his eyes were forever straying to the front door. Caroline saw the anxiety there and smiled reassuringly at him. Julie was being chatted up by a young man in a white T-shirt, but she didn't seem very interested.

Roy Campbell walked in at about 1.15 and David almost breathed an audible sigh of relief.

The other man was in his middle forties, plump, short and at least two stone overweight. He wore a shiny corduroy jacket which was the only thing David could ever remember seeing him in the entire time he'd been a visitor to the pub. His hair was streaked with grey in places but still retained most of its black lustre.

'Afternoon, Roy,' said David, pulling him his usual pint of light and bitter. Campbell exchanged pleasantries with the other men at the bar, two or three of whom then decided to retire to a table in the corner for a game of dominoes. Campbell was left alone at the bar.

'What was the phone call for, Dave?' he asked.

'I need your help,' the pub owner confessed.

'*My* help?' said Campbell, sounding surprised. 'With what?'

'Are you still interested in outer space?'

'I still study the stars, yes. I saw a comet a week or two back,' he said, proudly.

115

'You haven't seen anything unusual lately, have you?'

'Like what?' Campbell wanted to know.

David shook his head. 'It doesn't matter,' he said, noticing that someone was approaching the bar with two empty glasses. He hastily refilled them then returned his attention to Campbell. 'Look, Roy, I don't want you to breathe a word of this to anyone, but,' he hesitated, still reluctant to say anything at first. 'Do you still belong to that organisation that researches UFOs and things like that?'

'Not anymore,' Campbell told him.

David exhaled deeply.

'But you still believe in them?'

'Of course I do, it would be naïve to think that mankind was the only form of intelligent life in the universe. There's too much we . . .'

David cut him short. 'We've seen a UFO,' he said. 'All of us, the whole family. As a matter of fact, the damn thing has been showing up quite regularly.'

'Is that what you wanted to talk to me about?' asked Campbell, his pint halted half-way between the bar top and his mouth.

David nodded. 'Yes, but not here. Come through into the back, will you?'

He lifted the bar flap and Campbell followed him through into the sitting room.

'Sit down, Roy,' David instructed.

'Have you reported the sighting?' asked Campbell, excitedly.

'About four times, but the police don't want to know. That's why I wanted to talk to you. I knew you wouldn't laugh at me. Besides, it's got to the stage where we're getting desperate.'

There was a long silence then David spoke again. 'Do you know of any incidents where UFOs have been responsible for someone's death?'

Campbell looked shocked. He shook his head slowly. 'Not *directly* responsible,' he said. 'They might have caused the odd heart attack in older people if they turned up unexpectedly but never . . . death. Why?'

David avoided the question. 'This rash,' he gently touched his own skin. 'That didn't appear until after we'd seen the spacecraft for the first time and my daughter, Michelle, she's lost her voice. It must be something to do with it.'

'When did you first see this thing?' Campbell wanted to know.

'About two weeks ago.' David went on to describe the glowing yellow object.

'Does it look the same every time you see it?'

'Basically, but sometimes it turns white, or looks as if it's spinning.'

'And you say the whole family have seen it?'

'Yes, and what's come out of it.'

Campbell was stunned. 'You haven't had a close encounter have you?' he asked, a vague smile on his lips.

'The ship, saucer whatever the hell you want to call it. It landed on the lawn at the back of the pub, some kind of creature got out. Caroline, Julie and the kids saw it.'

'My God,' murmured Campbell, excitedly. 'What did this creature look like?'

'I didn't see it myself and I think the others were too frightened to remember it properly.' He paused. 'Do you believe what I'm telling you, Roy?'

'Of course I do.'

'Well, there's something I think you ought to see.'

As David and Campbell walked towards the wood where the UFO had first landed, David explained everything about the events of the past two weeks, neglecting only to mention the visit from the men in black.

'It's unusual for one subject to be the focus of so many visitations,' said Campbell.

'Perhaps they like us,' David said, sardonically.

'It's also unusual for the visitations to be so . . . violent. Usually the encounters are harmless. I mean, I've heard of people being harmed by prolonged exposure to a UFO. There was a boy in Wales, back in 1975 this was, he saw a

UFO and he suffered from epileptic fits and temporary blindness afterwards. But that sort of case is very rare.'

'Well, Caroline is convinced that this thing, whatever it is, is trying to kill us,' David said.

'That's unlikely, Dave,' said Campbell.

'You haven't seen it.'

David led the way into the wood, the first time he'd entered it for over a week and, immediately, he felt a strange lethargy begin to wrap itself around him. His stride seemed to shorten and slow up and his body began to feel heavy. As they drew nearer the landing site, he could hardly move his feet. Campbell, however, did not appear to notice any difference and, for the first time, David began to wonder if he was imagining the condition. Campbell had told him that UFOs sometimes had the effect of hypnotising those who saw them. Had he himself been mesmerised into feeling the way he did by repeated exposure to the glowing yellow object? Was it all in his mind? He tried to lift his leg and found that the effort was almost beyond him. He couldn't be imagining this. Could he? As the two of them reached the circular patch of scorched earth, David stood still. He felt as if he had run a hundred miles, every muscle in his body ached.

'There,' he said, pointing towards the patch of burnt ground.

Campbell ran an appraising eye over the damage, marvelling at the sight of the destruction and the size of the circle. It looked as if someone had turned a flame-thrower on the place. David watched him as he moved cautiously across the burned area, squatting to inspect the charred ground. David himself felt dizzy, as if he'd drunk too much. The trees and bushes around him seemed to swim into a blur then clarify once more. He shook his head as if trying to dispel the mist which was clouding his brain.

'Well, what do you think?' he said to Campbell who was inspecting a lump of blackened wood.

'It could have been a craft of some kind,' he said. 'On the other hand, it might just have been some space debris or, I did tell you I saw a comet a while back.'

'Then where's the crater?' David demanded.

'But Dave, you said when it landed on your lawn, it left no trace at all. No marks, nothing. Why should it leave traces here and nowhere else?'

'How the hell should I know?' David rasped. 'You're supposed to be the expert.'

'What about the footprints you said you found?' asked Campbell.

David lowered his head slightly.

'I haven't been able to find them since that night,' he confessed.

Campbell nodded. 'Did you say that Michelle had been here?' he asked.

'Yes, why?'

'The area might be mildly radioactive, that could be why *she's* not feeling so good.'

'But Dean was here too, he's fine.'

Campbell stood with his hands on his hips surveying the blackened ruins before him. David struggled to move, as if some invisible force had him in an unbreakable grip and was squeezing tighter. Finally he could stand it no longer.

'Damn it,' he rasped.

Campbell spun round.

'Can't you feel it?' David asked him.

'Feel what?'

'There's some kind of force here.'

'Probably an electro-static field. That could be why your clocks and watches stop every time the craft appears. It's bound to be stronger here if this is where it landed.'

David felt his entire body stiffen, the hairs on the back of his neck rise. He felt as if he had stared at some kind of gorgon and his limbs were slowly turning to stone. He finally closed his eyes tight shut and managed to turn away. As he tore his gaze from the patch of incinerated ground he felt some of the feeling returning to his limbs. His fingertips tingled and he flexed them. It felt as if he had pins and needles.

He began walking back, out of the wood.

Campbell followed. 'I don't know what to do, Dave,' he said. 'I've never heard of anything like this before. Anywhere.'

'Do you think they're trying to kill us?' David asked.

'Why should they want to?'

'We don't even know what they are, I don't think it's time to be questioning their motives.'

The two men reached the outskirts of the wood and, as they emerged into the open once more, David felt as if a huge weight had been lifted from his shoulders. He stamped on the ground as if trying to restore circulation to his legs.

'I don't know how much more of it the rest of the family can take,' he told Campbell. 'Caroline's talking about leaving.'

'What do you think?'

David shook his head. 'No way. I intend to see this thing through to the end. It's just that I feel as if I've got no one to help me. And I don't want you to mention any of this to anyone, please Roy.'

'I wish there was something I could do,' said Campbell.

'There is. Keep your eyes peeled when you're looking through that telescope of yours. I wish someone else *could* see the bloody thing, I'm beginning to wonder if we *are* all imagining it.'

'Why don't you contact a UFO group like NUFON,' Campbell suggested.

'What the hell good would that do?' David wanted to know.

'They'd send investigators. They'd check out your story, do tests and . . .'

David cut him short. 'No. I don't want word of this getting out, I told you.'

'Then how the hell can you expect anyone to help you?' Campbell said in frustration.

David didn't answer, he was looking up at the sky.

It was nearly 2.45 p.m. by the time the pub was emptied. Campbell had gone about half an hour before that and now, as they washed glasses, David stood in the bar and told Julie and Caroline everything that the amateur ufologist had said. Dean and Michelle, as well as the two dogs, were outside.

'So,' David said, finally. 'It looks like we're on our own.'

'We should get out of here, David, even if it's only for a few days,' Caroline said.

'What's that going to solve?' he wanted to know.

'At least we wouldn't be in danger,' she said.

He was about to speak when the phone rang. David picked it up.

'Hello, The Horseshoes public house, can I help you?'

Silence.

'Hello, The Horseshoes public . . .'

'You were told to remain silent, Mr Ellis.'

David shivered. He stiffened.

'Who is it, David?' Caroline asked.

He waved a hand for her to be quiet.

'You were warned. This is not the concern of others.'

He recognised the voice only too well. It belonged to one of the men in black.

'Now listen to me you bastard . . .' David began angrily but his words seemed to fade away.

'Do not speak to anyone else about what has gone on.'

David swallowed hard.

The line went dead.

He held the receiver away from his ear, staring at it for a moment before gently replacing it on the cradle.

'It was one of those men, wasn't it?' said Caroline.

He didn't answer.

'Wasn't it?'

David nodded.

'What did he say?'

'Another warning for us to keep quiet.'

'What did Roy Campbell say about those men?' Caroline wanted to know.

'I didn't tell him,' David confessed.

The phone rang again and David hesitated, his hand hovering over the receiver. He looked across at the two women, both of whom were looking intently at him. There were worried expressions on both of their faces.

David paused a moment longer, the strident ringing seemingly growing louder, then he snatched up the receiver.

'Yes,' he said.

'Dave?'

'Who is this?' he asked.

'Roy Campbell. What's up?'

David relaxed. 'Nothing. Sorry, Roy. What can I do for you?'

Campbell chatted away and Caroline saw the expression on David's face darken as he listened, only saying 'yes' or 'no' every so often. Finally, he merely grunted and put the phone down. When he turned around his face was as white as a sheet.

'David. What's the matter?' Caroline asked him.

'That was Roy Campbell. He wanted to know if we'd seen any strangers lately. Anyone driving a big flash car.'

Caroline froze as he continued.

'It seems he had visitors. Just before he got home his wife told him that two men had been looking for him. They asked for him by name. They drove off in a big black car. She said that the peculiar thing about them was, they were both dressed like undertakers. All in black.'

Fourteen

August 12th

There was something missing from the kitchen but David couldn't quite put his finger on what it was. He prodded at his food, eating small mouthfuls, finally pushing the plate to one side, deciding to pour himself another cup of tea instead. He had a slight headache but he felt sure it would pass. He had heard nothing more from Roy Campbell or from anyone else, fortunately. He glanced across at the phone as if expecting it to ring at any minute. Answering the damn thing had become something of a test of nerve just lately, never knowing who was going to be on the other end.

He looked around at the rest of his family. Caroline was glancing at the newspaper and munching a piece of toast. Julie was standing over by the work-top preparing the pastry for the pies and pasties. Dean was tucking merrily into his bacon and eggs. The things which had happened didn't seem to have affected him in any way. Not on the surface at any rate. But, David thought, who knew what thoughts and fears fermented inside that ever questing ten-year-old mind?

Michelle, still unable to speak, sat next to him. She looked a little pale, but, if anything, that paleness only made her look more beautiful. When would her voice come back? What could he do about it? The poor little devil looked tired, though. They all did. Everything, on the face of it, *seemed* to be more or less all right, apart from Michelle's voice. Why then, David asked himself, did he feel so uneasy? There was something not quite right here and trying to figure out what it was . . .

One look at his plate told him.

There were some pieces of bacon still on it. Usually King and Sabre were seated beside the table ready to devour any scraps given to them. But there was no sign of either of the Alsatians. Curious, thought David. Come to think of it, he hadn't seen them since he got up. They hadn't been in the kitchen when he'd walked in over ninety minutes ago. It was now almost 8.30 a.m. and he'd seen neither hide nor hair of them.

They must be in the garden, he reasoned. But that in itself was unusual. They were always around at meal times, hungry for the scraps which the kids sometimes slipped them, despite having been told on numerous occasions not to.

He got to his feet and walked across to the kitchen window, peering out into the garden beyond.

No sign of them out there.

'Is something wrong, love?' asked Caroline as he made his way into the sitting room and looked out of the bay window.

'No, nothing,' he said, scanning the area outside for the dogs.

David bit his lip contemplatively. 'Where are the dogs?' he asked, matter-of-factly.

'You let them out this morning didn't you?' said Caroline. 'You were up before me.'

'They weren't in the kitchen when I came in. I thought you'd let them out.'

The kitchen door was still locked.

David and Caroline exchanged brief but telling glances.

'Maybe they've run off into the woods, Dad,' Dean suggested. He hadn't seen David carefully unlock the door.

'Probably,' said David, trying to smile.

'They might be chasing a rabbit,' the boy added.

David stepped out into the garden where he was joined, a moment later, by Caroline.

'Where the hell could they be?' he muttered.

'Well, if they're not in the house, they must be out here somewhere,' said Caroline, realizing how ridiculously obvious her statement was.

'We'd better find them,' said David. 'You stay here, the kids will help me look.' He touched her gently on the cheek. Dean and Michelle joined him, and the little trio set about discovering the whereabouts of the missing animals. But, even as they began their search, David could not suppress the strange sense of foreboding which filled him.

The dogs weren't in the fields to the south of the pub. They weren't on the hillsides around it. They didn't come when they were called. They didn't even bark.

David walked as far as the end of the driveway which opened out onto the main road into Matlock but they were nowhere in sight. He made his way quickly back up to the pub where the two children were looking in the garden again. Whistles and calls filled the morning air and David knew that if the dogs were within earshot they would have come running.

Michelle was beginning to look upset and she gazed up at David with tear-filled questioning eyes. Dean too was becoming tired of the search.

'Oh, where are they, Dad?' he said, wearily.

David could only shake his head.

'It's like they've just disappeared into thin air.'

David was thinking exactly the same thing. 'They'll come back when they get hungry,' he said, trying to reassure the children, but failing miserably. One dog going missing he could perhaps understand. But both of them? And where could they have gone? He glanced at his watch and saw that it was almost opening time.

'You two carry on looking for them,' he told the kids. He ruffled their hair. 'They'll be back, don't worry.' He smiled thinly and headed back into the pub, but before he did he cast a quick and anxious eye over the sky.

The dogs had still not returned by noon. One or both of them usually wandered around the public bar during opening hours, but, as David spoke distractedly with his regulars, the animals were conspicuous by their absence. Every so often

David would scan the bar in the hope of catching sight of one of the dogs but his vigil went unrewarded. As far as he knew, Dean and Michelle were still out looking for the missing Alsatians and he was sure that, as soon as they were found, he would be told about it. But one thought nagged persistently at the back of his mind. What if they weren't found? He tried to qualify his thought. They couldn't simply have disappeared. It wasn't possible. He felt uneasy. A lot of things had happened lately that didn't seem possible, was this just the latest chapter in the story? No, David refused to believe it. The dogs would turn up.

He tried to forget about the incident and chatted as amicably as he was able to his customers.

Caroline, bringing a fresh batch of hot pies from the kitchen, was also worried about the disappearance of the two dogs. She knew that the children had returned over thirty minutes ago, still having found no trace of the animals, but she didn't intend telling David just yet. She too found it hard to comprehend how two fully grown Alsatians could simply vanish, especially when the pub had been securely locked up. There was no possible way they could have got into the cellar, simple logic said that they were outside, but where?

David pulled the bolt across the door and exhaled deeply. The last customer had just left, somewhat the worse for drink, and the pub owner had been glad to see the back of him. He hadn't recognised the man and that fact alone had made David wary of him. In the past week or so, they had all come to distrust strangers.

Julie and Caroline were busy washing and drying glasses, ready for the next session in less than two-and-a-half hours. Michelle and Dean wandered into the bar looking somewhat distraught and weary.

'We can't find them, Dad,' said Dean, answering David's unspoken question. 'We even went over to the farm but no one there had seen them.'

David nodded slowly. 'I can't understand it,' he said, as if finally admitting defeat. There was nowhere left to search now. The dogs were gone for good, by the looks of it. The children returned to the sitting room and sat down, the sunshine outside not acting as an enticement to them as it usually did. They slumped in armchairs and stared into empty air. The three adults exchanged puzzled glances then went on with their own tasks feeling helpless and somewhat uncomfortable. David decided to check the cellar. One of the barrels had run dry. He would need to replace it before they opened for business again. He unlocked the cellar door and reached for the light switch.

Nothing happened when he flicked it.

'Damn,' he muttered. The bulb must have blown. He went to the kitchen and retrieved another bulb then headed back towards the open cellar door.

There was a loud cacophony of barks and snarls and something came skidding from the blackness of the cellar.

First one dog then another, they hurtled into the bar as if flung by some invisible force behind them. David and the two women looked on in amazement, and even the children heard the commotion and came running in from the sitting room.

Sabre and King stood in the middle of the bar, the hair on the backs of their necks prickled, their ears pointed forward. As David moved towards them, Sabre began growling loudly. His lips slid back to reveal long teeth and David stopped in his tracks.

'What's wrong with their fur?' asked Julie, pointing at the animals.

All eyes turned on the animals, but, more correctly, to the dark patches which covered their normally sleek fur. They looked from a distance like burns. The fur seemed to have been scorched in some way.

'They were in the cellar,' said David, softly, edging his way around the dogs which were still growling.

'But the door was locked, there's no way they could have got down there,' Caroline protested.

'Easy, boy,' David cooed, moving closer, his hand

outstretched.

King barked and snapped his jaws together. David hastily withdrew his hand, pausing when he saw the dog gathering itself together, hind legs taut as if it were preparing to spring at him. Dean stepped forward towards Sabre.

'Keep away from him,' David hissed, his eyes never leaving the two Alsatians which by now had stopped growling and were content to bark at their cautious master. The last time David could remember seeing them like this was when the spacecraft had been hovering over the pub.

They were obviously frightened and David continued his advance cautiously, watching them intently for signs of any movement but the dogs remained still and, as he drew closer, he could see that both were shaking.

'Come on,' he said softly. 'Good boy.'

He extended a hand once more and this time was relieved when neither of the dogs snapped at it. Even their barking began to subside. The snarling anger and fear slowly became whimpering bewilderment. David managed to get within two feet of King, the dog now lowering its head slightly and watching his owner as he approached. Sabre pawed the floor and looked away from David who was now within arm's length. He swallowed hard, if the dogs suddenly decided to turn on him now he would have little or no chance to defend himself. The breath stuck in his throat, his mouth was dry.

'Easy. Easy,' he said softly and reached out those last few inches until his hand was on King's back. He began to stroke the dog, looking at the singed patches of fur as he did so. The other dog nuzzled against him and he looked up and waved the rest of the family forward. Dean began stroking Sabre, the dog turning and licking his face.

Caroline merely stood gazing at the tableau before her.

'How could they have got into the cellar?' she said again.

David could only shake his head. He had no answer for her.

'They couldn't have been down there since this morning,' she persisted.

As he stroked King, David felt the rough patches of

scorched fur and he inspected it more closely. The patches were mainly along the dog's back and sides, the pattern repeated with Sabre. They looked like trainee circus performers who had yet to master the art of jumping through a fiery hoop. But, other than the burns, the dogs seemed relatively unscathed and, most peculiar of all, they looked as if they had just been given a bath. The areas which remained unsinged were silky smooth, the fur like silk to the touch. Surely if they had been outside for any length of time, David reasoned, or, even down in the cellar (which wasn't the cleanest of places), they would be slightly dirty, carrying perhaps the odd traces of mud. But there was no sign of dirt, earth or grass. Not even dust which, had they been in the cellar, they could not have failed to have picked up on their coats.

David chewed his bottom lip and continued stroking the animal, ensuring that they were both calmed down. Dean and Michelle were smiling happily as they fussed over Sabre who reciprocated by licking both of their faces. The kids giggled and David too, smiled. Only when he looked at Caroline and Julie did his expression change to one of bewildered foreboding.

He sucked in a troubled breath. Could the dogs have been down in the cellar all day? Surely there was no possible way they could simply have been . . . taken. Was that the word? He wondered how much of this was the paranoia taking over again. What did he think had happened to the dogs? Had they been teleported away by some unseen force? Beam me up, Scotty, he thought and tried to smile but it dissolved into a sneer. There was no way they could have been in the cellar all day. It was as simple as that. *Something* had happened to them, but God alone knew what it was. David looked at the burns once more. He touched one of the patches tentatively and King growled throatily. It was only the fur which was singed, the flesh beneath appeared to be untouched. The force of heat which had marked the dogs seemed to have been controlled rather than random. Something which made David feel even more uneasy.

Eventually, when he was satisfied that both of them were stable once more, David got to his feet and told the children to feed the dogs. Dean and Michelle led them into the kitchen, where, helped by Julie, they filled the animals' bowls with food.

'What could have happened to them, David?' Caroline asked.

David could only shake his head.

'Perhaps the cellar door *wasn't* shut,' he said, realizing that he was indulging in that time-honoured pastime of clutching at straws.

'Do you really believe that?' Caroline asked him.

'I don't know what to believe,' he hissed. 'I'd like to know what the hell those burns are on their coats.'

It was her turn to be silent.

David exhaled deeply. 'No, I *don't* believe they were in the cellar all day,' he said, wearily. 'For one thing, if they had been, they would have barked, tried to get out.'

'Then what?' Caroline wanted to know.

'I wish I knew.'

'It must be something to do with what's been happening lately,' she said.

David didn't answer.

'Perhaps they were showing us.'

'Showing us what?' he wanted to know.

'How easy it would be for them to take one of us if they wanted to,' Caroline told him.

'No, that's ridiculous,' he said, dismissing the idea. 'We don't know *what* happened to the dogs. There's nothing to say that the . . . the . . .' He seemed reluctant to say the word. 'Spaceship had anything to do with their disappearance. I mean, for one thing, we haven't seen it around here today have we?'

'Just because we haven't seen it doesn't mean it's not around,' she said.

'Oh come on, now, love. The bloody thing has been all too willing to show up in the past, why should it stop now?'

Caroline didn't answer. She walked across to him and put

her arms around him. He squeezed her tightly to him and felt her body shaking.

'I'm so frightened, David,' she said, her voice cracking.

'It'll be OK,' he whispered.

He wished he could believe his own words.

Fifteen

August 13th

'David.'

He heard his name being called, felt a hand on his shoulder.

'David.'

He woke up with a start, almost cracking his head on Caroline's. She was bending over him, holding a cup of hot milk. He took a few seconds to re-orientate himself then he rubbed his eyes and sat up. He was sitting in his armchair in front of the TV. It was 12.15 a.m. according to the clock on the mantelpiece.

'I must have dropped off,' he said, almost apologetically.

'Yes, you did, as soon as you sat down,' Caroline told him.

'Sorry,' he said and sipped his milk.

It had been busy that evening, all three of them had been on the go since opening time. A coach party had stopped off at the pub, about thirty of them. They'd monopolised the lounge bar but David hadn't cared, they'd also drunk a vast amount between them. Some more than others. The driver had finally managed to herd them all together at about 11.00. Like some kind of demented shepherd he had driven his merry and, in some cases, completely inebriated, flock back to the coach and driven away.

'Where's Julie?' asked David.

'She went to bed,' Caroline told him. 'I think I'll go up now.'

David looked vague. He gazed at Caroline but seemed not to see her.

'David, are you all right?' she asked.

'What?'

'Are you all right?'

'Oh, yes, just tired,' he told her. 'I'll be up as soon as I've finished this.' He held up the cup of hot milk.

She leant forward and kissed him on the cheek but he seemed not to realize what she was doing. Caroline studied his profile for a moment then headed for the stairs. David sat back in his chair, massaging the bridge of his nose between his thumb and forefinger. He could feel a headache beginning. The TV was still on and he glanced at the screen. There was some large lump of what looked like mouldy jelly in the process of devouring an old man. David laughed to himself as he watched then the caption card came up signalling the arrival of the adverts. *The Blob*, it announced and then the screen was filled by the sight of a woman washing nappies. David got up and switched the set off. He unplugged it then sat down again to drink his milk.

Outside, rain had begun to fall lightly, a light breeze throwing the droplets against windows to form erratic rhythms.

David went to the kitchen and checked on the dogs.

Both were sleeping in their respective baskets. He checked that the back door was locked and bolted then moved through into the bar to repeat the procedure there.

Caroline undressed and slipped into bed. She too felt tired and was pleased to sink into the welcome oblivion of sleep. It was not long in coming.

Across the landing Julie slept fitfully.

Satisfied that all the doors were secured, David returned to the sitting room and planted himself in his armchair to finish off his milk. God, he felt so tired. He could hardly keep his eyes open. The cup seemed to blur before him, but something kept him conscious. A noise he'd first detected a matter of minutes ago.

A low buzzing sound which seemed to grow in intensity.

David tried to listen against the persistent rhythm of the rain. There was no doubt about it, the buzzing was getting louder. He tried to rise but found that he couldn't. It felt as if someone had nailed him to the chair. He could only stare down helplessly at his body which remained immobile. His eyelids were getting heavier.

The buzzing grew louder.

Suddenly, the TV screen exploded into a vivid profusion of colours. Zig-zags and broad slashes of colour ran rampant across it and loud blasts of static came from the sound box. It was like watching an electronic firework display as the colours were gradually displaced, bleeding away into stark black and white. One part of David's mind told him that he was seeing things, that he had gone insane, even.

The TV was still off and unplugged.

Yet the blinding flashes of interference went on and, try as he might, David could not raise himself from the chair.

The buzzing sound grew louder and now, out of his eye corner he saw something else. A bright light, tinged with yellow, but he couldn't turn his head to get a proper look. His eyelids continued to close, as if some maniacal puppet master had attached strings to them and was pulling them down.

Everything went black.

David found that he could open his eyes but, on doing so, he wished he hadn't. Blinding white light poured through the bay window, hardly deflected by the curtains. He walked across to the window and looked out, his limbs now finding the strength they had previously lacked.

The spacecraft had landed, as before, on the lawn at the back of the pub. It stood motionless, a throbbing yellow-white ovoid which gave off the most powerful luminosity David had ever seen. He tore his gaze from it, fearing he might be blinded. As he reeled away from the window he saw that the clock on the mantelpiece had stopped. Likewise his watch. He lurched drunkenly towards the door of the bar and stumbled through as he heard the first of many powerful thuds on the front door. For long seconds he stood there, just watching the door, listening to the slow steady pounding. But then he felt as if he must cross to it and as he walked, his legs lost that leaden feel. It was as if he were walking on the moon, his steps weightless and easy.

The pounding stopped as he reached the door, his hand hovering over one of the bolts.

He felt a force unlike anything he'd experienced before which seemed to be physically pulling his hand towards the bolt and, against his will, he found his fingers closing over it, pulling it back.

David unlocked the door and stepped back, the leaden feeling returning to his limbs. His head felt as if it weighed a ton and his chin dropped forward to rest on his chest. Once more his eyelids began to droop.

The door swung open.

David opened his mouth to scream but no sound would come forth.

Something large moved past him and he tried to see what it was. The creature didn't appear to have any discernible shape but it possessed what passed as arms and legs. The feet, however, appeared to be about six inches above the ground. The intruder seemed to be floating.

David could not see because he was unable to turn, but, nevertheless, he was able to sense that the creature was heading for the staircase.

Once more David tried to scream but found that he couldn't.

The light.

It was the same as before.

Caroline opened her eyes a fraction, trying to prepare herself for the onslaught of brightness when it eventually came. She was floating on air again, a warm breeze beneath her back. And she was naked.

She tried to turn her head, to see the invisible wires which kept her suspended in the air and which kept her arms pinned down. She couldn't move, then, once again, as she had done the first time, she felt her arms being grasped. Cold hands fastened around her wrists and raised her arms until they were level with the rest of her body. She opened her mouth and attempted to call for help but she was unable to make a sound.

Then another pair of hands gripped her ankles and she felt her legs being parted. Wider and wider until it seemed she would be torn in half. She tried to jerk her head up to see who was doing this to her but she couldn't move.

Her heart almost missed a beat as she felt a burning sensation just above her navel. As before it spread slowly to her thighs and pelvis then upward to her breasts. There was no pain, just an almost unbearable heat and she felt that she would explode if it did not soon stop. Her eyelids closed, shutting out the bright light and she could only wait as she felt something cold touch her face. Something which reminded her of overused soap. A sticky, clinging sensation which sent waves of revulsion through her.

And then she felt the burning sensation in her throat.

Help me. Help me. But the words were formed only in her mind and, even though her lips fluttered, there was no sound.

David tried to move his feet but couldn't. No more than he could lift his head to see what was going on. All he knew was a being of some kind had entered the pub and was now, as far as he knew, upstairs. Upstairs with his wife and children. The thought made him tremble and he tried even harder to move, but it was useless. He opened his mouth and let out a despairing moan. His eyelids closed again and, once more, everything was in darkness.

Sixteen

David awoke with a start and found himself sitting in the armchair. It was light, the early morning sunshine was piercing several chinks in the curtain. David sat bolt upright, his hands immediately going to his neck which ached terribly. For what seemed like an eternity he sat there, aware of how cold it was in the room. He got to his feet, anxious to discover why it felt so chilly. David crossed to the window and pulled back the curtains, shielding his eyes against the bright sunlight which streamed in, but despite the sun it still felt cold.

A newspaper which lay beside his chair stirred in the breeze and David noticed that it seemed to be coming from the direction of the bar. He headed for the connecting door and walked through.

The front door was wide open.

For a moment, David stood staring at it, his mouth open in disbelief.

He remembered locking it the night before. He remembered sliding the bolts into place. He remembered tugging on the handle to see that it was secure. And yet now it yawned open, allowing the cold breeze into the pub.

Who had opened it, he wondered?

He bowed his head. Why was he finding it so hard to think? His brain was fogged, incapable, it seemed of coherent reasoning. Unable even to remember what had happened the previous evening. And how come he had fallen asleep in the chair? He never did that normally. He tried desperately to think back to the night before, but the recollections eluded him.

He crossed to the door and closed it, leaning against it for a moment, shaking his head. Why couldn't he remember anything?

He stood still for long moments until not one but two loud screams galvanized him into action.

David turned and raced for the stairs.

He found Caroline and Julie together in one bedroom. Both were clearly distraught, Caroline sitting on the side of her sister's bed.

'What's wrong?' asked David, his face etched with fear.

The women paused, as if waiting for the other to speak first, but it was Caroline who broke the silence.

'The nightmares,' she said. 'We both had the same nightmare again.'

'Only this time it was much more vivid,' Julie added.

David walked into the room and sat down on the chair by the dressing table, looking at the two women.

'What happened?' he asked.

One at a time they explained their nightmares. The lights. The sensation of floating. The cold hands on their legs and wrists. And the burning sensation.

'I had a dream too,' said David, when they'd finished.

He told them the details of his own nightmare.

Neither of them said anything.

'I've read of two people sharing the same experiences before,' David said. 'Especially people close, like you two. But never three people. It's like we all shared a part of the same dream.'

The two women looked carefully at him as he continued.

'But there's something else. When I woke up this morning, the front door was open. Just the way I'd dreamt I opened it, but I can't remember anything after that.' All he'd been able to recollect of the dream was opening the door and seeing the light. Always the light.

His face and neck were once more blotchy and red. It appeared that the rash had returned too.

'Is it possible that we could all have dreamt the same

thing?' Julie asked.

'It's possible,' said David. 'But I should think the chances are about ten million to one.'

Caroline shook her head. 'It seemed so real,' she said softly.

Julie pulled herself upright in bed, wincing at a slight pain in her lower abdomen.

'Are you all right?' asked David, noticing her momentary distress.

She nodded.

'What exactly happened at the end of the dream?' he asked her.

'The same as before,' she answered. 'I was naked, I felt something pushing my legs apart and then this feeling of heat just here,' she pressed the area she meant, letting out a small yelp of pain as she did so. The colour drained from her face as she cautiously prodded the indicated spot. It felt very tender around her navel, exactly where the pain had begun in the dream. David and Caroline looked on in concern. Julie gently pulled her nightdress open, careful not to expose too much of herself, her eyes straying to the area around her navel.

It was all she could manage to choke back the scream.

All three of them saw it.

Just above her navel was a large, dark triangular scar, the tip of which reached up nearly as far as her breasts.

'Oh my God,' she murmured, her hands beginning to shake.

David studied the scar with horrified bewilderment, almost unaware that Caroline was also pulling open her own dressing gown. She lay back on the bed and murmured something under her breath.

Like Julie, she bore a triangular scar just above her navel.

'A dream?' she said, cryptically.

David tried to think of some way of explaining the marks. There was none. Some kind of psychosomatic reaction to the nightmares perhaps? Auto-suggestion, whatever the hell that was, he'd heard the word used on a documentary about the supernatural some months earlier. Was it that? He looked again at the two marks. If they had been measured they could not have been placed with such perfect matching similarity.

'David, we couldn't have been dreaming,' said Caroline. 'Not all three of us, and there's no way to explain these marks.'

'If we weren't dreaming, what are you trying to say?' he wanted to know.

She paused for a moment. 'The craft landed once before, outside. The night you drove into Matlock to fetch the police. Something, an alien, got out of it that night. It tried to break in and now you said that when you woke up this morning the door was open. *You* remember opening it last night.'

'I still don't see what you're driving at.'

'Julie and I dreamt that we were being experimented on,' said Caroline, flatly. 'And now we find these.' She pointed to first her scar and then her sister's.

'You think you were taken aboard that spaceship and experimented on?' said David, somewhat incredulously. If not for the circumstances, he might have found it difficult to stop himself laughing.

'I think that's the only answer,' Caroline said. She was adamant.

'Julie, what about you?' David said.

'I don't know what to think,' she confessed. 'All I know is, whatever happened to us, to all three of us, wasn't natural, and I'm sure it wasn't a dream. I think Caroline's right. How else could you explain these scars?'

He had no explanation. Something beyond comprehension was almost literally staring him in the face he knew he could seek no refuge in logic. Logic. The word was practically a redundant one in his vocabulary after the events of the past few weeks. But his wife and sister-in-law abducted by aliens? Even after everything that had happened, David could not bring himself to accept that fact. Nonetheless, he still had nagging doubts about what he himself had witnessed the previous night. Things were still hazy, half-remembered images which seemed to form no coherent whole. He *did* remember opening the front door. But why? And the light. He remembered the light. Everything else was still vague, like looking through memories from behind a gauze veil.

'Are the kids OK?' he asked.

'I haven't checked on them yet,' Caroline admitted.

David got to his feet. 'Perhaps they had dreams too,' he said. 'If we all had the same dream . . .' The sentence trailed off. He wandered across the landing to Dean's room and went inside. The ten-year-old was huddled up beneath the blankets, just the top of his head sticking out. He didn't move as David entered the room.

'Dean,' he called. 'Are you awake, son?'

The little boy turned over and looked up as his father approached the bed.

'Dad, I heard a noise,' he said. 'I heard someone scream. What's happening?' He sat up slightly.

David perched on the edge of his bed and ruffled the boy's hair.

'Are you OK?' he asked.

Dean nodded. 'What was that noise, Dad?' the lad persisted. 'Is it one of those spaceships again?'

David froze, his body stiffening. 'No,' he said, swiftly. 'Dean, did you dream last night? Can you remember?'

'No, Dad, I don't think I did.'

'And you didn't hear any noises, or see anything during the night?'

'Like what?' the boy wanted to know.

David shrugged and tried to smile but it looked false and he could sense it.

'I heard some noises, I wondered if you might have,' he said, trying his best to avoid the question.

Dean shook his head.

He watched as his father got up and left the room, making his way along the landing to Michelle's room. He knocked and walked in, knowing that the poor little devil wasn't going to be able to tell him anything anyway. He paused as he entered the room, surprised to see that she was not in bed but standing at her window looking out on to the lawn beyond. She didn't turn as she heard the door open but merely kept on gazing out into the garden. David crossed to her and put an arm around her shoulder. He kissed the top of her head and looked down, surprised to see a drawing pad and a couple of pencils perched on the wide window ledge. As he looked, she hastily covered the sketch pad with one thin arm.

'Are you feeling OK?' he asked her.

She nodded.

He guided her over to the bed, noticing that she clutched the sketch pad to her chest as he did so.

'Have you been up long, sweetheart?' he wanted to know.

She nodded again. David could see the dark rings under her eyes. The poor little mite didn't look as if she'd got much sleep.

'You didn't dream last night did you, Michelle?'

She shook her head.

He noticed how tightly she gripped the sketch pad to her.

'Can I see what you've been drawing?'

She shook her head.

'Is it a secret, then?'

She bowed her head, her body quivering, and only then did he see tears beginning to drip on to the paper. David pulled her to him.

'What's wrong, love?' he asked, his voice heavy with concern. She threw her arms around him and, as she did so, the pad fell to the floor. David glanced down at it, the breath catching in his throat.

Michelle had always been a good little artist, from the earliest age, and now she liked nothing better than to sit and draw while her brother fought new wars with his model soldiers. Her representations were usually remarkably accurate. That was one of the reasons why David felt his hand shaking as he reached for the pad.

On it, in bold strokes, was a perfect sketch of the spacecraft. And something else. Something which looked shapeless yet had appendages which could pass for arms and legs. And it was floating above the ground, she had shaded in an area beneath it to emphasise this. And now the memories came hurtling back into David's mind like an out of control steam train. He felt the hairs rise on the back of his neck. He *knew* that shapeless thing. He had seen it.

'Did you imagine this, Michelle?' he asked, holding up the pad.

She shook her head.

'You actually saw it?'

A nod.

'Last night?'

A nod.

'Write down what happened for me, sweetheart,' he instructed, giving her a pencil, sitting with his arm around her shoulders as she wrote, her hand shaking slightly:

I heard this noise last night and I woke up and I looked out of the window and the light was there again. It was the light that we seen before and I saw the space ship that we saw. A man got out and he looked horrible not really like a man. he looked like a blob with arms

David read as she wrote.

'Is that the man you drew?' he asked her.

She nodded then continued writing:

I was very fritened and I wanted to cry but I watched the man. He banged on the door and he came in and I got very hot I couldnt sleep I was so scared. But he made me cry.

'Did you actually see him when he was inside the house?' David wanted to know.

She shook her head.

The pencil hovered over the pad once more.

then he went back to the flying sawser and it went away but I didnt go to sleep because I was two fritened.

He hugged her once more, brushing a tear from her cheek.

'It's OK now, sweetheart,' he said, holding her tight for long moments.

His mind was full of thoughts. Thoughts which he scarcely dare to entertain. If Michelle was telling the truth (and, he reasoned, how could she possibly be lying?) then what happened to Caroline and Julie? Experimented on by aliens from outer space? My God, it sounded like the title of some tacky science fiction film. *The Inter-Galactic Surgeon*, he mused, humourlessly. *Marcus Welby, MD* (Martian Doctor, of course). *Emergency Ward 2001*. The jokes did not seem funny in the light of what had happened. Yet everything David had ever believed in was threatened should he accept totally whatever it was that happened the previous night. He must retain some of his cynicism or he feared he would go mad. Yet he felt now that was impossible.

He stayed with Michelle for a long time, re-reading what she'd written, looking at her drawings, and all the time he felt a knot of fear growing bigger and bigger in the pit of his stomach.

So far, but for the scars which the two women bore and Michelle's loss of voice, no one had suffered any physical damage from the repeated visitations. He wondered how much longer it would remain so. Were they, as Caroline had insisted, in real danger?

He dared not contemplate what might happen next and, as he hugged Michelle closer to him he realized just how helpless he was. There seemed to be nothing he could do. He couldn't fight these invaders, he didn't even see them most of the time. He took the pad from his daughter and ripped off the piece of paper which bore the writing and sketches.

'Can I keep this, love?' he asked.

She nodded vigorously, as if she were pleased to see the back of it.

David folded it up and put it in his pocket.

'Are you all right now?' he asked, smiling when she nodded, even managing a little smile.

He left, and only when he had his back to her did he allow the expression of fear to cover his face once more.

A dream? It didn't look like it. He swallowed hard and made his way downstairs. He found Julie and Caroline already in the kitchen, sitting over a cup of tea, now both dressed.

Sabre and King were padding around the table looking up to see if any biscuits might be tossed their way.

David sat down and gratefully accepted the cup of tea which was pushed his way.

He clasped his hands around the cup and drank.

'So, what do we do now?' Caroline wanted to know. 'Stay here and wait until they come back for us all?'

David didn't answer. He merely reached into his pocket and pulled out the piece of paper which he'd taken from Michelle's room. He handed it to Caroline who read it, her expression of weariness changing to one of outright horror. She handed it to Julie.

'She saw it all,' said David.

'Then it wasn't a dream?' Caroline said. There was a long pause. 'David, we can't stay here any longer. How is Michelle ever going to get her voice back? We don't know if she'll ever speak again! We've got to leave before something happens, before they kill one of us.'

'And where do you suggest we go?' he said.

'My parents in Nottingham.'

'And what about the pub?' he asked. 'Who takes care of that?'

'David, for God's sake,' Caroline gasped. 'Our lives might be in danger and all you're worried about is the pub.'

He was about to answer when the phone rang.

For long moments all three of them sat looking at each other then David finally got to his feet and picked up the receiver.

The line went dead immediately.

He held the receiver away from his ear and looked at it irritably for a second before replacing it on the cradle.

'What's your answer then, David?' Caroline asked him. 'Do we leave or not, because I'm telling you, if you won't come then *I'll* take the kids.'

'I said we weren't leaving here, no matter what happened.'

'Circumstances were different then. We've got no choice. You can call Stewart and Rita Poole, they'll take over for us while we're away.'

David sighed. 'How long were you thinking of going for?' he asked.

'I don't know,' Caroline confessed. 'Until this blows over.'

'But we don't know how long that will be.'

'I don't *care* how long it is,' she protested. 'I just want to get away from this place.'

'Assuming we go to your parents, what do we tell them? "Sorry to bother you but we've been having a spot of trouble with UFOs lately and we were wondering if you could put us up until it's all over?" They'd think we were mad.'

'We'll tell them that we decided to have a break, that we thought it would do the kids good to have a break. Something like that.'

David lowered his head slightly, contemplating the problem.

The phone rang again and he muttered to himself as he got to his feet.

Julie and Caroline glanced at each other as he raised the receiver to his ear.

'Hello, The Horseshoes public house.'

Silence.

'Hello.'

'Mr Ellis.'

David visibly stiffened as he heard the voice, his hand closing around the receiver until his knuckles turned white.

'What do you want?' he asked, his voice low and trembling.

'You know who we are then?' the voice asked.

'Yes,' he snarled, recognising the metallic sound of the voice. It was unmistakably that of one of the mysterious men in black.

'Your daughter is a very gifted girl,' said the voice.

'What the hell do you know about my daughter, you bastard?'

Caroline crossed to the phone, seeing the angry expression on David's face. The colour had drained from his cheeks, the knot of muscles at the side of his jaw pulsed angrily.

'Have you told anyone yet of what happened last night?' the voice asked.

'What if I have?' he barked.

'You were warned to be silent.'

'If I ever see you again, I swear to Christ I'll kill you.'

'Take care of your daughter, Mr Ellis,' the voice said, softly.

The line went dead.

David put down the receiver, his eyes on the piece of paper which Michelle had written and drawn on. He swallowed hard, and as he brushed the hair from his forehead, his hand was shaking. He looked at Caroline.

'Call your parents,' he said. 'Tell them we'll be there later today.'

Seventeen

August 16th

It was not the way of Bill Grant or his wife Ivy to ask questions of their children. They did not need to ask to know that something had been very wrong at the pub. Three days ago they had received a phone call from Caroline asking if it would be all right if the whole family came over for a few days and, without asking any questions, they had agreed. Bill and Ivy lived in Nottingham, in a house which was much too large for their needs. It contained five bedrooms and three bathrooms, for one thing, but Bill, who had once owned a hosiery firm in nearby Mansfield, had been able to retire at fifty-nine and he felt that the house, despite its size, was a justifiable monument to his lifetime's toil and hard work. Besides, they had a lot of friends and at least one of the spare bedrooms was usually occupied at weekends. Even the dogs were welcomed. It was grand for them, having two of their three daughters home again albeit for reasons which neither of them felt like discussing. But Ivy didn't probe, not even when she had them alone. Just as Bill did not question David as the two of them walked around the large garden enjoying the sun, listening to the excited chatter of the children from next door who had come over to play with Dean and Michelle. There were five of them in the garden, peering at the fish in the pond, kicking a ball around on the expansive and well kept lawn which Bill still maintained himself. No, they asked no questions. If anyone wanted to tell them the truth then they'd be happy to listen but they weren't stupid enough to think that the entire family would leave Matlock and the pub for 'a little break' unless there was a very good reason. They hadn't pressed it when Caroline had dismissed Michelle's inability to speak as the after-effects of laryngitis. Nor had they asked why David was suffering from such a nasty rash. Though in the past three days it had all but disappeared.

The pub itself had been left in the capable hands of Stewart and Rita Poole, good friends of the Ellises who had worked as relief bar-keepers for them on previous occasions.

Now, as David and Bill stood in the garden, beneath the blazing sunshine, drinks in their hands and the dinner cooking inside, the dogs and kids enjoying themselves, David thought how distant all their problems had been. But the pub never left his mind. It had left its mark on them all in more ways than one. Even now he found himself shifting uncomfortably every time the phone rang and, as expected, each time it had been unjustified. He had rung The Horseshoes the day before to ask if everything was running smoothly, half hoping that Stewart Poole would say no. 'Sorry to tell you this, David, but we've been terrorised by repeated UFO sightings and also, shapeless blobs have been floating through the house leaving triangular burns on my wife's stomach.' As far as David could tell, none of the phenomena had been repeated. Stewart would have mentioned something, he knew it. Obviously things *were* running smoothly. Perhaps, he wondered, hardly daring to entertain the thought, it was all over.

Bill was talking about his flowers. He took great pride in his garden, but David found that the words went in one ear and out the other, a fact not lost on his father-in-law.

'Something on your mind, Dave?' he said, smiling.

'Sorry, Bill, I was miles away,' he apologised.

Bill Grant smiled and sipped at his drink. 'I noticed.'

'Bill, there's something I need to tell you,' David began. 'You'll probably think I'm a nutcase but I might as well say something anyway.'

'I was wondering how long it would take you,' said Bill.

David looked puzzled. 'Dave, I'm not blind. Nor is Ivy and neither of us is stupid,' said Bill, smiling. 'We knew there was something going on the minute Caroline phoned.'

'Parents' intuition?'

'If you want to call it that.'

'Do you believe in the existence of UFOs?'

Bill raised an eyebrow. 'I think it would be naïve to dismiss them as hoaxes. Too many people see them every year. It's the same with ghosts. There's too much evidence available for people to ignore them.'

'Then you *do* believe?'

Bill nodded.

'That's why we left the pub for a while,' David told him. 'I haven't said anything until now because it hasn't seemed like the right time but, well, I'm sick of bottling it up. The rash that I had, Michelle's throat complaint, they're because of close contact with a UFO. We've seen one up close, it chased the car. It even landed in our garden.'

David looked up to see that his father-in-law was listening intently to every word.

'How long has this been going on?' Bill wanted to know.

'Weeks.'

'Didn't you get help?'

'The police didn't want to know. It seems we're the only ones who saw it. Even the couple who are in charge of the pub now haven't seen anything unusual.' He went on to explain a few of the other incidents which had happened, even the strange meeting and subsequent phone conversations with the men in black.

'Christ,' said Bill, when he'd finished. 'No wonder you wanted to get away.'

'It's up to you if you want to tell Ivy,' David said. 'Caroline and Julie might have done it already.'

Bill nodded.

'Since we've been here,' David told him. 'The incidents seem to have stopped. It's been three days now since anything happened. Perhaps it is all over.'

'How long are you planning to stay here?' Bill asked.

'I don't know. It was Caroline's idea to come in the first place. She was worried in case something happened to the kids.'

Bill laid a reassuring hand on his shoulder.

'You stay as long as you need to,' he said. Both of them began walking back up towards the house, the smell of cooking food wafting invitingly on the breeze as they drew nearer. David paused to kick the ball around with the kids and Bill watched in amusement as Dean tried to take the ball from his father. It bounced clear towards Michelle who ran after it, competing for it with King who was leaping around like a puppy. She picked it up, smiling broadly.

'I've got it.'

David froze, his eyes locked on his daughter.

She had spoken. The words not croaking whispers but full throated and clear.

Michelle even seemed surprised herself.

Even Dean was staring at his sister in shocked bewilderment.

'My voice . . .' she began, touching her throat. It was as if something had been unlocked inside her. Caroline emerged from the kitchen, tears welling up in her eye corners.

'Mum, listen, I can talk again,' said Michelle, beaming.

Caroline ran to her and embraced her. David held both of them and found that even he was weeping softly.

A friend of Ivy's, a doctor who lived two doors away, came round for his usual cup of afternoon tea as he did every Sunday and examined Michelle while he was there. If it had been laryngitis the infection was now gone, he assured them. She should have no problems from now on. He also took a look at David's rash, although there was hardly anything to see by this time. He left somewhat bewildered, but with a piece of Ivy's home-made gâteau to pacify him.

Caroline had wondered whether she should mention the scars which she and Julie bore, but in the end had decided not to bother. That was one thing which David had left out whilst telling Bill what had happened.

Besides, next to Michelle's recovery, everything else seemed unimportant. Caroline congratulated herself on having decided to leave the pub, for only in the past few days had things begun to change. David was back to his usual jolly self and now Michelle . . . Everything seemed to be rosy.

David phoned the pub again that night but Stewart Poole had nothing to tell him other than business was booming.

David asked if anything unusual had happened.

Stewart had sounded puzzled but David wouldn't be more specific. Yes, something unusual *had* happened.

David had frozen when he'd heard that. He'd asked what it was.

Stewart told him that the unusual thing had been that Reg Wheeler, usually a beer drinker, had switched to vodkas and had to be carried home.

David had laughed until tears ran down his cheeks, and most of that was due to relief.

That evening they had all sat outside on the patio until ten o'clock when the sun finally began to turn from gold to crimson to purple and finally the sky was dark but for the smattering of bright stars. The kids were tucked up safely in bed, the five adults sat in the pleasant night air drinking and talking. David found that he could once more enjoy gazing at the sky and find beauty there, not the fear and anxiety which he had come to know. Even the sight of a bright red light did not make him jump because he realized immediately that it was an aeroplane. He felt more relaxed than he could remember for many years, let alone weeks. Caroline squeezed his hand gently and he smiled at her.

They retired at around 11.30, David sinking into the comfort of the bed, watching as his wife undressed. He sat up in surprise as he saw her reflection in the mirror on the wardrobe door.

Up until now, even Caroline herself had not noticed but David pointed it out.

'The scar,' he said. 'It's gone.'

Caroline inspected the area where it had been and found that there was just a white patch of skin. Nothing else. Even as she stood before the mirror, there was a soft knock on the bedroom door, then Julie entered.

She saw Caroline inspecting the area where the scar had been and she smiled.

'So yours has gone too, has it?' she said and they began to laugh like a couple of schoolgirls.

Eighteen

September 5th

It was time to return. David had decided the night before. They had been away from the pub for nearly three weeks, the kids should be back in school, they had to return and he felt that the time was right. He and Caroline had talked it over and she had agreed with little reservation. He had expected protests from her but she agreed readily, as did the rest of the family. He had rung Stewart Poole at The Horseshoes every other night and on no occasion had there been anything strange to hear. By all accounts, things were almost welcomingly boring. Business was steady, Poole had told him, nothing spectacular though. David didn't care about that. Everything seemed to have returned to normal, that was all that mattered. There was a time when David wondered if Stewart had not mentioned anything over the phone for fear of worrying the family. A brief but telling image of returning to be told a catalogue of weird happenings crossed David's mind but he rapidly dismissed it. Stewart Poole was not the kind of man who kept things of that magnitude to himself. If *he'd* seen a UFO David and half of Matlock would know about it by now. No, the pub owner was satisfied that it was safe for them to go home.

Now he and the rest of the family stood beside the Volvo, beneath the hot sun in the driveway of Bill and Ivy's house.

'Are you sure you've got everything?' said Ivy, fussing as mothers are prone to do no matter what the age of their offspring.

Caroline assured her that they had.

'I don't know how to thank you, Bill,' David said, shaking hands with his father-in-law.

Bill waved it away and embraced David. 'Take care,' he said. 'If you ever need any help again, you know where we are.'

David nodded.

Goodbyes were exchanged and Ivy rebuked herself for allowing a tear to fall. She kissed the kids, she kissed her daughters, she kissed David. She almost kissed the dogs. The family were finally settled in the Volvo and David started the engine. He wound down his window to let in some fresh air then drove off.

Within ten minutes they had left the bustling city behind. Another five and the countryside began to undergo a transformation. Houses were replaced by tree-lined lanes, tarmac became grass and traffic was reduced to only a handful of vehicles.

High above them, the sun beat down, growing hotter as the day progressed. By noon it was well into the seventies and David pulled over into a lay-by where they all piled out of the car to eat the sandwiches Ivy had prepared for them and, in Dean's case, to scurry off behind the nearest bush to relieve himself.

They ate at a leisurely pace, enjoying the bright, warm weather. David glanced up at the sky which was clear and as blue as sapphires. It looked like being a beautiful day.

'How far to go, Dad?' asked Michelle.

David smiled, so pleased to hear her voice again after what had seemed like an eternity. Caroline saw his smile and reached across to grip his free hand.

'We'll be home in about ten minutes, sweetheart,' he told his daughter.

King barked.

'Yes, that's right,' said David. 'Ten minutes.'

The family burst into fits of laughter.

David noticed how relaxed they all seemed, the kids chattered for most of the journey and Julie did her best to supply them with answers to their innumerable questions, but now, as he swung the car onto the road which led up to the pub itself, a strange and expectant hush fell over them. When anyone spoke now, it was quietly, as if they were in church, and all eyes were glued to the pub.

There were a number of cars parked outside, and the place looked immaculate in the sunlight which glinted off its white walls. He slowed down and glanced at his watch. It was just after 1.15 p.m. David parked the Volvo in its customary place then set about removing the cases from the boot. The rest of the family piled out of the car, the dogs bounding towards the building as if they couldn't wait to get back inside. That in itself seemed to reassure the others for it was as if a collective sigh of relief had been released. The children dashed off after King and Sabre while Julie and Caroline helped David with the cases. The pub seemed to hold a sense of tranquillity which David had not noticed for a long time. He admitted to himself that he'd felt somewhat apprehensive as they'd approached but now he strode purposefully towards the building as if he were greeting an old friend.

Stewart and Rita Poole continued serving for the rest of the day and night, refusing David's offer to help them. When, that night, the last customer had left for home and the doors had been bolted, Stewart, Rita, David, Caroline and Julie all sat in the bar and shared a bottle of wine. David thanked them for their help then drove them home.

On the way back he stopped the Volvo close to the range of low hills which masked the wood where he'd discovered the craft's first landing site. David got out of the car and walked slowly to the top of the hill, savouring the clean fresh air. The smell of grass mingled with the faint aroma of wild flowers which grew in abundance on the slopes. When he reached the top he paused, looking down at the wood below. It seemed harmless now. Nothing more than a few trees and bushes. He felt none of the menace which he'd experienced before. The night sky was clear, a full moon shining down. There was a slight breeze but the night was mild. The entire scene looked so serene it was difficult for him to imagine that anything so mysterious and terrifying could have happened there such a short time ago. Below, on his left, was the pub, its outside lights still burning. He smiled with pride, happy to be back, but as he wandered slowly along the hill top, scanning the heavens, he could not resist a sigh of relief. It really did look as if things had returned to normal. He stood still gazing up at the sky, looking at the silvery pinpricks of stars which littered the canopy of blackness. David inhaled another lungful of that clean air and smiled contentedly. Another five minutes and he made his way back to the car and drove the remaining few hundred yards home.

He found Caroline sitting up in bed when he finally made his way upstairs. She was brushing her hair. David undressed and slid into bed beside her, a broad smile on his face.

'What's tickling you?' she asked, grinning.

'Nothing in particular,' he said, his smile still as broad as a Cheshire cat.

She nudged him in the ribs and he laughed.

Neither Caroline or any of the others had mentioned anything about what had gone before. Whether they had forgotten all about it (something which he doubted) or whether they had just decided not to mention it he wasn't sure. Nevertheless, no one had mentioned lights in the sky, aliens or UFOs and, as far as David was concerned, that was fine.

Caroline finished brushing her hair, laid the brush on the dressing table and flicked off the bed-side light. She moved closer to David and he put his arms around her. She was asleep within minutes, he lay awake, watching the curtains gently waving, coaxed by the breeze. Then, he drifted off into a peaceful and untroubled sleep. One thought remained until he nodded off.

It was finally over.

He went to sleep with a smile on his face.

Nineteen

September 6th

The day passed quickly, or so it seemed to David. Business was brisk and it didn't seem five minutes between the time he opened up and the moment when he saw the final customer out and locked the door. The warm weather had persisted and he was pleased when Caroline announced that she wanted to join him when he walked the dogs that night. Julie volunteered to tidy up alone while they went out.

They left by the back door, King and Sabre bounding ahead of them chasing a small red ball which David threw for them. He and Caroline walked slowly, arm in arm, allowing the dogs to run ahead of them.

'It's like we were courting again,' he said, smiling.

'Did we do things like this when we were courting?' she asked, as they began to climb the hills near the pub, the slope rising gently to its flat top. 'You were too busy trying to get me in the back seat of your car.' She giggled.

'What a cheek,' said David, in mock indignation. 'My intentions were purely platonic, my motives honourable.' Then, he too laughed.

Caroline looked up at the sky, scanning the clear purple-hued heavens.

She opened her mouth to say something but decided not to. She merely squeezed his arm tighter. They reached the top of the hill, watching as the dogs bounced around like puppies as they searched for the ball which David had thrown. It had rolled towards the copse at the bottom of the hill. David smiled as he watched them, but the smile froze as he saw King stiffen and begin growling. Sabre followed his example.

Both of them were facing the woods.

David pulled loose and began his journey down the slope. 'Stay there, love,' he said. 'I'll be back in a minute.'

As he neared the bottom of the slope both dogs began barking at something which, as yet, David couldn't see. He felt a powerful feeling of unease close around him like an invisible glove. The trees seemed to lower down at him challengingly and he felt a bead of perspiration pop on to his forehead. The dogs were still barking but not moving forward.

David paused and scanned the area in front of him. A breeze rustled the leaves, the sound reminding him of croaking whispers.

He saw something move about two feet ahead of him.

It was a grass snake. The reptile slithered hurriedly away, the dogs still barking at it, but no sooner had it disappeared beneath a bush than they seemed to forget it and turned their attention to the ball once again. David swallowed hard and smiled to himself. Come on, no more paranoia, he rebuked himself. Nevertheless, he couldn't resist a furtive glance amongst the trees closest to him. Their conspiratorial whispers now sounded only like sighs. He turned and headed back up the hill to where Caroline waited. They began walking again.

Julie had just finished putting the last glass away when the phone rang.

She crossed to it and picked up the receiver, wondering who could be calling at such a late hour.

'Horseshoes pub, can I help you?' she said, as brightly as her tiredness would allow.

Silence. Then a hiss of static.

'Hello?'

She heard breathing at the other end and almost laughed.

Not a dirty phone call, she thought, grinning. Not at this time of night.

'Who is this?' she said.

The voice, when it came, was low, metallic sounding. 'Welcome home.'

The line went dead.

'Hello, hello.' Julie flicked the cradle. She gently replaced the receiver, a cold chill running over her skin. She looked towards the front window, and felt a sudden compulsion to gaze out at the night sky. There were only clouds.

Julie looked at the phone, as if expecting it to burst into life again at any moment and, as she did, she caught sight of the clock above the bar.

It had stopped, the hands frozen at 11.56

She checked her watch.

That too, had stopped.

It was then that she noticed the faint buzzing sound.

'You know,' said Caroline. 'I think this is the most beautiful time of the day.'

'Night,' said David.

She playfully thumped him. 'You know what I mean. It's so peaceful.'

He nodded in agreement. It was, indeed, peaceful. 'Well, peaceful or not,' he said. 'We'd better get back.' He looked at his watch and found that it had stopped. 'Damn,' he muttered and asked Caroline what the time was.

Her watch had also stopped.

David's face lost a little of its colour. 'Let's get back.'

They both turned and with the dogs racing on ahead made their way back towards the pub. As they reached the top of the slope which would take them down into the hollow where the building lay, they both heard an all too familiar sound. A loud buzzing, like that of many bees, growing in intensity.

David found his breath coming in gasps and, almost involuntarily, both of them now found that they were running.

They almost fell at the bottom of the hill, scrambling on to reach the back door. The dogs bolted inside, David and Caroline followed. Julie stood in the sitting room, facing the bay window, her eyes fixed on something outside. She was pale, her features waxen. David and Caroline joined her, anxious to see what she was looking at.

'Oh my God,' Caroline whispered.

High up in the night sky, moving slowly back and forth, was the glowing yellow shape. The burning ovoid which, they could see, was rapidly descending.

'No,' Caroline gasped and a large tear ran down her cheek. 'Why? Why has it come back?'

No one answered her question.

The buzzing grew louder and all three of them became aware of the heat which was filling the room as surely as invisible gas. It seemed to be sucking the oxygen from the air. The yellow egg swooped lower, and as it did so there was an explosion from behind them as the sitting room light bulb shattered. It was the first of many, as the strip lights in the kitchen disintegrated, then the lights in the bar itself.

The glowing craft came lower.

It began to circle the house, slowly at first but then with increasing speed until it was moving faster than a whirlwind. The buzzing sound grew in pitch until it was a deafening whine.

All three of them spun round as they heard what sounded like the impact of a hammer on stone.

A crack the length of a man's arm appeared in the sitting room wall, then another. And another. Pieces of plaster fell to the floor.

Outside the craft gained speed, moving so fast it was little more than a blur.

It grew hotter inside the building.

Another crack was made, this time in the ceiling. Then the plaster groaned once more and split as another part of the wall was riven.

'It's creating a vacuum,' David yelled. 'It could tear the building apart.' But all they could do was crouch helplessly on the floor as pieces of masonry fell from the walls and ceiling. In the bar one of the bottles exploded on its optic, showering the place with liquor and lumps of glass.

Michelle and Dean appeared in the doorway and Caroline rushed across to them, pulling them to the floor with her. They were both crying. The dogs had stopped barking and were whimpering like puppies, the hair on their coats raised high by the electro-static charge which was being created.

And still the craft continued its mad, circular manoeuvre as if it wanted to tug the building up from the ground into the sky itself. More cracks appeared in the plaster. One of the panes of glass in the latticed bay window splintered.

Caroline screamed.

The buzzing became unbearable and David thought that his ear-drums must burst, but the one part of his mind which was still functioning rationally told him something. Something which made *him* want to scream out in terror and frustration. He realized now that they had been singled out. *His* family. There had been no activity for over three weeks, because whatever was up there had been waiting. It had waited for them.

For them alone.

They had been chosen.

But even as he tried to digest that terrible thought, the buzzing seemed to diminish, the craft slowed its breakneck circling until it just hovered in the sky. Then, slowly, it rose, higher and higher. Still visible but growing smaller until it was just a little larger than the stars around it.

Then, it was gone.

David got slowly to his feet and checked that everyone else was all right. No one spoke, and only the whimpering of the dogs and the soft sobs of the children broke the silence. He crossed to the window and looked out.

There was no sign of the craft. The blazing yellow shape was nowhere in sight. He lowered his head, wiping perspiration from his brow with one shaking hand. Broken glass and pieces of plaster lay everywhere, some crunching beneath his feet as he walked. He wandered into the bar to inspect the damage there. Three bottles had exploded, the steady plink-plonk of dripping liquid the only sound in there.

He poured himself a large scotch and downed it in one, allowing it to burn his throat and stomach.

They'd thought it was over. They'd prayed that it was over.

David looked towards the window, towards the boundless canopy of darkness above.

It had waited for them.

The words echoed inside his tortured mind.

They had been chosen.

It was then that he realized they could never remain in the pub. They weren't leaving, they were fleeing for their lives.

He put down the glass and went back into the sitting room.

Epilogue

The Ellis family left their home and did not return. They moved to another part of the country in the hope of escaping the force which seemed to have singled them out for visitation. The Horseshoes public house was taken over by new owners and, to this day, no extra-terrestrial activity has been reported from that location.

Exactly why the Ellises were singled out for such a terrifying ordeal will never be known. The fact is, they *were* singled out and every incident set down on the preceding pages actually happened. To coin a cliché, truth once more proved itself to be stranger, and in this case more terrifying, than fiction.

The family now live in welcome anonymity, untroubled by any more encounters such as they experienced in 1981. What is set down in these pages remains the only testament to the curious happenings. The veracity of that testament will be questioned by many who have read this book, and it is your right, as readers and thinking people, to question what you have read. The choice, ultimately, is yours. You either believe or disbelieve. It would be much easier to be sceptical, to dismiss the entire episode as fiction but it would be both unjust and also naïve to do so. The facts remain unaltered. If we choose to ignore the truth perhaps we do so at our own peril. There are thousands of UFO sightings every year in Great Britain alone, most of which can be explained away by logical argument and evidence.

This is not so with the experiences of the Ellis family.

Of the plethora of strange events which have been described perhaps one of the most curious is the persistent activities of the so called Men in Black. The Ellises are not the only ones to have encountered these strange beings, whoever or whatever they are. Other people have been approached by them. 'Silenced' by them. Warned by them. They always seem to appear after a UFO sighting as was the case with David Ellis and his family. In 1967, Colonel George P. Freeman of the United States Air Force spoke of such 'silencing' of witnesses. A Canadian UFO witness, Carmen Cuneo, was told, in 1976, by a mysterious visitor that he should refrain from telling his story for fear of retribution. And, a year later, a Mexican, Carlos Del Los Santos, was stopped in the street by three of the men in black and warned not to appear on the TV show he was due to speak on concerning his own UFO experiences.

Regarding Caroline and Julie's 'dreams' of being abducted and experimented on, here a potentially logical explanation for what happened founders eventually against a barrier of the unexplained. It would not, at first glance, be difficult to dismiss the 'dreams' as just that, especially after what had already happened to them. On a psychological level, the dreams of being experimented on are not uncommon. The doctor remains a powerful figure in human consciousness and both women had recently been in contact with one, namely Dr Fenwick who examined Michelle. But how can one explain the scars which they found on their abdomens? As with so many of the other incidents, science *has* no answer.

So, once more, it is the prerogative of you, as readers, to believe one way or the other. The facts have been presented. The truth has been told. And, despite those who may cast scorn on this chronicle, *truth* is what it is. Nothing can alter that. The Ellis family were subjected to a series of events which defy logic and understanding but which, nevertheless, happened.

For them, the ordeal is over.

For others, it may just be beginning.

Author's note

Unlike the Coombs family, whose story was widely reported in both local and national newspapers in 1977 and who eventually became the subject of *The Uninvited*, the Ellis family and those connected with this true story, preferred to remain anonymous. Therefore, I have respected their wishes and changed the names of those involved. That is all that has been changed. The incidents, incredible as they seem, are all too real.

Frank Taylor.

STAR BOOKS BESTSELLERS

THRILLERS

SHATTERED	John Farris	£1.50*	☐
BLOODSPORT	Henry Denker	£1.75*	☐
THE AIRLINE PIRATES	John Gardner	£1.25	☐
THE INFILTRATOR	Michael Hughes	£1.60	☐
IKON	Graham Masterton	£2.50*	☐
TERROR OF THE TRIADS	Sean O'Callaghan	£1.50	☐
HUNTED	Jeremy Scott	£1.50	☐
DIRTY HARRY	Philip Rock	£1.25*	☐
MAGNUM FORCE	Mel Valley	£1.50*	☐

WAR

BLAZE OF GLORY	Michael Carreck	£1.80	☐
CONVOY OF STEEL	Wolf Kruger	£1.80	☐
BLOOD AND HONOUR	Wolf Kruger	£1.80	☐
PANZER GRENADIERS	Heinrich Conrad Muller	£1.95*	☐
THE RAID	Julian Romanes	£1.80*	☐
GUNSHIPS: NEEDLEPOINT	Jack Hamilton Teed	£1.95	☐
THE SKY IS BURNING	D. Mark Carter	£1.60	☐
TASK FORCE BATTALION	Tom Lambert	£1.60	☐

STAR Books are obtainable from many booksellers and newsagents. If you have any difficulty tick the titles you want and fill in the form below.

Name _____

Address _____

_____

Send to: Star Books Cash Sales, P.O. Box 11, Falmouth, Cornwall. TR10 9EN.

Please send a cheque or postal order to the value of the cover price plus:
UK: 45p for the first book, 20p for the second book and 14p for each additional book ordered to the maximum charge of £1.63.

BFPO and EIRE: 45p for the first book, 20p for the second book, 14p per copy for the next 7 books, thereafter 8p per book.

OVERSEAS: 75p for the first book and 21p per copy for each additional book.

While every effort is made to keep prices low, it is sometimes necessary to increase prices at short notice. Star Books reserve the right to show new retail prices on covers which may differ from those advertised in the text or elsewhere.

*NOT FOR SALE IN CANADA